MARK PARENT

SPIRIT SCAPES

MAPPING THE SPIRITUAL & SCIENTIFIC TERRAIN AT THE DAWN OF THE NEW MILLENNIUM

Northstone

Editors: Michael Schwartzentruber, Dianne Greenslade
Cover artwork: Scott Waters
Cover and interior design: Margaret Kyle
Consulting art director: Robert MacDonald

Northstone Publishing Inc. is an employee-owned company, committed to
caring for the environment and all creation. Northstone recycles, reuses and composts, and encourages
readers to do the same. Resources are printed on recycled paper and more environmentally friendly
groundwood papers (newsprint), whenever possible.
The trees used are replaced through donations to the Scoutrees for Canada Program.
Ten percent of all profit is donated to charitable organizations.

Canadian Cataloguing in Publication Data
Parent, Mark, 1954 –
SpiritScapes
Includes bibliographical references and index.
ISBN 1-896836-11-9
1. Christianity and culture. 2. Church and the world. I. Title
BR115.C8P37 1998   231.7   C97–911087-4

Published by Northstone Publishing Inc.
Kelowna, British Columbia, Canada

Printing
9 8 7 6 5 4 3 2 1

Printed in Canada by Transcontinental Printing Inc.
Louisville, Quebec

DEDICATION

This book is dedicated to the
Pereaux United Baptist Church
without whose support
it could not have been written.
And yet,
of which it may be said,
that if all churches were like it,
this book
would not need to have been written.

And to

George Rawlyk
mentor,
friend,
fellow Christian
"'Blessed are the dead who die in the Lord,...'
'Blessed indeed,' says the Spirit
'that they may rest from their labors,
for their deeds follow them!'" (Rev. 14:13)

CONTENTS

ACKNOWLEDGMENTS

I would like to thank, with deep love and appreciation, my wife Cathy and our three children, Jeremy, Meaghan, and Kaitlyn for their support in writing this book. Not only did they put up with many occasions when I was unavailable to them, but they also encouraged me to continue when I lacked the discipline to do so.

In addition, a debt of gratitude must be extended to Nova Morine. Nova volunteered, as a person interested in spirituality and faith, as well as one knowledgeable in the English language, to read through the chapters in their initial form and help make them more readable and interesting. As well, Brad Lockner, a long-time friend and author, heard of my efforts and made the same offer for all the chapters. To Brad and to Nova my heartfelt thanks. They made this book better, although they cannot be held accountable for my perspectives and conclusions, or for any errors.

Several other people helped with various sections of the book. To George and Valerie Lohnes, who came over late one night and provided me a forum to sort through some thoughts, sincere thanks. To Dan Gibson, who read the chapter on the revivalist churches such as the Toronto Airport Church (and felt that I was wrong in my openness to this movement), sincere thanks. To my parents, Hazen and Hazel, for Dad's love of history and Mom's love of language, as well as for helping recollect the events surrounding Che's death, sincere thanks. To countless friends and church members, too numerous to mention, who expressed interest and support, sincere thanks.

A final word of appreciation must be extended to Mike Schwartzentruber and all the staff at Northstone Publishing. To be able to work with such a sensitive and skilled group of people is a treat!

Nearing the end of the writing this book, I am reminded of a comment by the novelist Madeleine L'Engle who noted in her autobiographical trilogy *The Crosswicks Journal* that all creative acts run the risk of failure. And so, to the divine being whom I name as God, to the one who through the Spirit can turn all our failures into successes, I offer my final word of thanks and a song of praise. Over 40 years of living have convinced me that it is only as we wait on God, that we can renew our strength; we can mount up on wings as the eagles; we can run and not grow weary; we can walk and never faint.

INTRODUCTION

THE FRUSTRATION AND THE SEARCH

THE FRUSTRATION

Increasingly, I find the contours of the Christian faith, on which I have been raised and which I have adopted as my own, empty and sterile. This seems like a strange comment from one who makes his living as a Christian minister within a traditional Protestant denomination. I do not make it lightly nor do I make it to capture the fancy of the secularized media in North America today. I make such a statement with great sorrow and enormous frustration. Traditional Christianity and the Christian church have sheltered me and given me a way of looking at life and death that I cherish. Hymns and biblical passages that I have memorized over the years will always stay with me as a source of comfort and strength. Moreover, I have known and continue to know wonderful churches and Christians who can only be described by the word *saintly*. Nonetheless, something is wrong with the Christianity and the Christian church that I know.

I am not alone in either my frustration or my search. In the July 7, 1996, issue of *Macleans*, the editor, Robert Lewis, notes that "throughout society, there is a rising interest in spirituality and a remarkable level of faith in God, even as attendance at traditional places of worship declines."[1] This book is written for those who have been raised as children within traditional Christian structures and belief systems or who have been influenced by those structures and belief systems but who increasingly find only emptiness and lack of fulfillment.

Many who feel the way I do have stopped attending church; the gulf between what they find and what they hope to find has become too painful. Others, traditionalists at heart who do not give up on institutions lightly, hang on like the biblical Jacob, desperately seeking the blessing. Ask almost any minister or committed church worker and they will tell you, once you have gained their confidence, about their frustrations and doubts. When you get past the religious jargon with which the committed church worker or the typical minister too often surrounds him- or herself, there is an inner emptiness, a void, similar to what is faced by the non-Christian and the non-church member. The non-church member and the active church member are more alike than they are different. This is to be expected since faith is never a private affair but always has social ramifications. Whether the non-Christian in Western society acknowledges it or not, the Christian faith has given our society a large share of its purpose and meaning, its values and perspectives. The vacuum within the church results in a vacuum within society, and the vacuum within society simply serves to underscore the fact that the Christian church is too often part of the problem rather than part of the solution.

When I first became active in the church in the 1970s, I thought that it was simply a matter of minor changes that needed to be made. The Jesus People legacy affected my thinking and I reasoned that if only we could update the language and the central symbols of the Christian faith all would be well. Keep the Lord's Supper, I reasoned, but substitute pizza and pop for bread and wine. The decade of the 1970s was a time of liturgical experimentation, as well as more substantive experimentation in religious institutions and practices, and I was affected by it as much as the next person.

As I entered seminary training, I moved to a deeper theological analysis and more reasoned criticism. I graduated in 1979 and entered the decade of the 1980s enchanted with the idea that a vibrant evangelicalism could blow away the dust of a long-dead liberalism and breathe life back into the dry bones of the Christian Church in North America. Today I feel the urge to run to the Toronto Airport Christian Fellowship (usually referred to as the Toronto Airport Church and part of what has been called the "third-wave of Pentecostalism"[2]) and throw myself into the strange experiences occurring there. People bark like dogs; they roar like lions. More commonly, they fall on the floor and lay there overcome by the presence of the divine within them. It sounds bizarre and yet thousands upon thousands from across the English-speaking world have attended services at the Airport Church and have claimed that the experience was beneficial. Perhaps, I reason to myself, Pentecostalism will succeed where evangelicalism has failed.

And yet while enticed, I resist making the pilgrimage to Toronto because I know deep inside that the renewal of faith must rest not simply on a new sense of emotional contact with the divine (either the spirit within as the New Agers claim, or God without as more traditional Christian groups claim), but on a new way of thinking. In this I am encouraged by my doctoral supervisor, Douglas Hall, who in his book *Thinking the Faith*[3] made the point that Christianity in North America has been deficient in the arena of systematic thought. We have been activists, feeding the poor or converting the sinful, but unless there are reasonable and accessible philosophical and theological frameworks that undergird that activism, it will fade away or spin out of control.

I differ from Douglas Hall in the extent of the changes I believe are needed. I suspect that Christianity will need to undergo as significant a change today as when it broke from its original Jewish background, if it is to survive and sustain North American and, indeed, Western society. We need a new Paul, a new Augustine, a new Aquinas, a new Luther. This assertion may be influenced more than it should be by an apocalyptic mind set. Yet, in Western thought, time is pregnant with meaning. The onset of the year 2000 cannot be dismissed as inconsequential and unimportant. Moreover, the death of the modern age and the birth of what has been called the post-modern age marks a significant stage in the thinking of the West, which calls for dramatic changes in religious thinking.

THE SEARCH

As a society we are emerging from an unprecedented period of history in which the divine was totally excluded from ordinary life or confined to the role of the "Unmoved Mover" who, once things were set in motion, had no further involvement with the created order. While post-modernity offers an opportunity for the rebirth of religion, it cannot be merely an echo of the religious thought of the past. This will not satisfy the religious aspirations of the post-modern individual. To use an analogy, it is possible to believe in the magic of Christmas as an adult, and to experience a sense of wonder and awe as one did as a child, but the belief system on which that wonder and awe are based is different from that of the child. There is no return to the garden of innocence after the aggressive skepticism of the modern age. But there is a way forward.

I just said, for instance, that what we need is a new Paul, a new Augustine, a new Aquinas, a new Luther. This listing of thinkers is itself an example of the change that is necessary. Every thinker I listed is male. And while I do not believe that males and females think as differently as some would claim, nonetheless, the feminine viewpoint has been sadly understated in the history of Western theologies. Not that I find everything coming out of feminist circles appealing. The antics of Starhawk, the celebrity neo-pagan writer, sometimes seem to me like a shallow response to a deep problem. The anti-male bias within the works of such writers as Mary Daly[4] is, hopefully, a passing stage that will be exposed for the reverse discrimination that it too often is. And yet, the explosion of the feminine perspective has been one sign of hope as we grope for a new and better future.

It is my intention to look at some of the new movements such as feminism and, by sorting through the chaff, find the grains of new ways of thinking about faith which may well revitalize the human community, in general, and Christians, in particular. For new ways of thinking about the divine (as well as old ways which have been forgotten and ignored by our society), are surfacing all around us, like bubbles rising from some subterranean spring.

There are some, of course, who are threatened by these new ideas, who believe that any change will be a negative and frightening experience. I choose to believe the opposite. At age 40 I feel more hopeful than I did at age 20 because of these bubbles of new life, new ways of thinking about God, and new forms of spirituality.

What shape will this new hope for the future take? Among which group is it to be found? To be honest, I do not know. In Christ's time, very few people knew that anything dramatic was happening. Jesus of Nazareth died a criminal's death on a criminal's cross. His life and death were, in the context of the Roman Empire, nonevents. His resurrection was even more obscure and hidden. How then do we do what Jesus himself asked his followers to do in the biblical story? How do we discern the signs of the times?

There are nine new movements that I believe offer hope for the future of religious faith: alternative medicine, the Gaia theory, the new physics, the New Age movement, the increase in near-death experiences, revivalist Pentecostalism, fundamentalism, liberation theology, and religious feminism. It may well be that all of them prove to be blind alleys, that while we gaze at what is happening in the Romes of our day – Washington, London, New York, and Moscow – a baby who will change the course of the world is being born in some obscure village, some modern-day Bethlehem. Yet, we must start somewhere because the Church in Canada and the United States is declining numerically. For many this might seem to be a good thing. For myself, I prefer to see the new building erected before the old one is demolished.

In Canada, church attendance has declined dramatically, as regular weekly attendance has plummeted from 60 percent in 1957 to 32 percent in 1986. In the United States, the drop has been more gradual, from 57 percent in 1958 to 40 percent in 1994. What these simple statistics hide, however, is a graying church population, a shift from mainline to more conservative or charismatic churches, and the explosive growth of an interesting if somewhat bizarre mixture of religious traditions and practices.

This decline in attendance too often has meant that concern for the institutional structure and survival of the Christian church in North America and in Europe has predominated over concern for spirituality. The energy of many congregations focuses inward on various reorganization schemes or on building maintenance. Financially supporting full-time religious leadership has become a burden to many small churches in urban and rural settings alike. Even more troubling, however, at least from the viewpoint of the traditionalist, is that Christian faith and practice, whether liberal or evangelical, does not seem to attract the young. Nor is it the case that young people in North America have flocked to other world religions such as Buddhism or Hinduism. Instead, we see an eclectic blending of various religious traditions and faiths, an interfaith stew if you will, combining the best and often the worst of religion.

In this regard, the writings of New Testament scholars and thinkers who have highlighted the Jewishness of Jesus are important. Increasingly I believe that the New Covenant is of value precisely because it was and is in continuity with the Old Covenant. Thus, while I dwell on what is new, I remind myself that the new always contains the old, and the new that dominates usually dominates because it contains the best of the old.

THE ENDURANCE OF THE RELIGIOUS

One thing is clear. The questions of life and death will never go away. The deep haunting cry inside when gazing at the stars overhead cannot be silenced. We are, as the theologian Paul Tillich noted,[5] incurably religious. This has surprised the scientifically minded and the empirical rationalists of our day. (This in itself is surprising. How could educated people be so blind to one of the fundamental aspects of what it means to be a human being?) The title of Paul Davies' book *God and the New Physics*[6] says it all. The popularity of alternative medical treatments, the formulation of the Gaia hypothesis, the increasing reports of near-death experiences (NDEs), have all challenged the linear thinking of modern science and philosophy.

In 1995, the *New York Times* reported that a group of top-level philosophers, scientists, and thinkers from around the United States convened a special three-day meeting in New York City in order to combat this supposed "flight from science and reason." As Malcolm Browne reported,

...the meeting was organized as a call to arms. Defenders of scientific methodology were urged to counterattack against faith healing, astrology, religious fundamentalism and paranormal charlatanism...

Participants deplored what they see as a growing trend toward the exploitation of scientific ideas to attack science. They cited the physics of relativity and quantum mechanics as pillars of 20th-century thought that are sometimes distorted by critics of science into arguments that nothing in science is certain and that mystery and magic have an equal claim to belief.[7]

Leaving aside the reactionary tenor adopted by the participating scientists and thinkers, it is true that much in the new movements I have chosen to study is vague and unhelpful. I have no doubt that quackery abounds within the practice of alternative medicine, for instance, or that many of the New Age channelers make a good living from seducing the desperate and the curious.

It is not so much the individual movements that make an impression on me as the impressive consensus of opinion. The dualisms of faith versus religion and of intuition versus logic are crumbling. Institutions, which were cherished in the past, are exposed as naked in the present. Perhaps the starkest example of this is the situation of the Roman Catholic church in the sparsely populated province of Newfoundland. Once honored and respected, the church has had to endure several highly publicized cases of systemic sexual abuse by Roman Catholic clergy. As a priest once shared with me, where before he was proud to let people know he was a priest, now he was ashamed. The professions that interpreted and passed on truth to previous generations are shamed and humbled. Clergy have been long dethroned. Now the medical profession and scientists fight back, seeking to preserve their dominant role as interpreters of truth; but theirs is a losing battle. For many, these changes are disconcerting, as they once were for me. Lately, I have come to see them as necessary preconditions for the new to grow and to blossom.

NINE MOVEMENTS

The first section of this book, that of new movements in the world of science, is one which may seem out of place in an analysis and presentation of new trends in religious thought and practice. We are conditioned to think that science is science, religion is religion, and never the twain shall meet. Many of us are amazed to read comments such as can be found in Frank Tipler's book *The Physics of Immortality* when, after arguing that science proves the existence of an afterlife, he concludes "science can now offer precisely the consolations in facing death that religion once offered. *Religion is now part of science.*" [emphasis mine][8] Whether science has subsumed religion or religion is in the process of subsuming science, it is clear that for a growing number of people the "two-book theory" of nature that delineated roles for science and faith is unsatisfying. Such people are not content with the little saying I grew up with, that science explains "how" and religion explains "why." Instead, new ways of thinking which put science and religion in dialogue and at times confrontation with each other are being explored.

This is, perhaps, the most exciting of the three sections in the book, but it is also the most difficult to understand for the scientific layperson. The world of alternative or complementary medicine appears opaque to many because we have grown up with a mechanistic view of the human body, where we view ourselves as flesh-and-bone machines, composed of separate parts which fit together to form the whole, rather than as vibrating energy fields. This latter is the stance taken by the new "medical" techniques.

Alternative medical practices confront us with the question, then, are human beings best thought of as biochemical machines, or as resonating energy fields? The weight of traditional medicine is against the latter interpretation and in support of the former; but an impressive and growing body of information suggests that, in some instances, alternative medical treatments based on a view of the human being as an energy field work better than traditional practices.

Within the first section we will also examine the Gaia hypothesis that the world is a living being and that we humans are often little more than parasites living on the skin of Gaia. What are the religious implications of this dethronement of the human creature? How does the Goddess Earth relate to the God Yahweh?

The final movement within this first section on science concerns the new thinking in physics. The philosophical and theological implications of the new physics still have not infiltrated our lives although, in the form of computers, the practical implications certainly have. Do Einstein's two theories of special and general relativity have anything to say to the world of religion? How does quantum physics change the way we understand our world? Is it, as Fritjof Capra claimed in his trend-setting book entitled *The Tao of Physics*,[9] a new way of reiterating truths which were discovered long ago in Eastern religious perspectives? Or are the discoveries of Einstein and Niels Bohr simply extensions of the thinking of Isaac Newton that was instrumental in the construction of the two-book theory?

The second section of the book, and undoubtedly the most bizarre, is one I have chosen to label "New Movements in Spirituality." The media have popularized these new movements in the past ten years, starting with New Age thinking in the 1980s and moving on to the new forms of Pentecostalism such as the Vineyard movement in the 1990s. It has been an uneven but an extensive treatment. By and large, the media has been sympathetic to New Age practices and teachings, while skeptical and questioning about the Pentecostal movement and the high profile Toronto Airport Church. Both movements, however, are more alike than they are different. Both are fascinated with alternative forms of medical healing. Both gather energy and attention out of the claim that we are living in a new age, the Age of Aquarius or the Age of the Spirit. Both claim justification for the truth of their perspectives from the changes in science described in section one.

Near-death experiences round out the study of new movements in spirituality. Originally, I planned to include NDEs under the heading of new movements in science, but further reading and reflection convinced me that writers in this field are not constructing an alternative scientific world view, but rather are challenging traditional spirituality. This is especially evident in the different constructs of salvation and of the afterlife, which prevail in near-death visions. They depart substantially from the traditional views.

I have included the third section, new movements in theology, because traditional Christian theology also has been part of this reorientation of spiritual vision. Three movements of particular interest and importance, particularly in regard to their social vision, are fundamentalism, liberation theology, and religious feminism. While many would agree with the inclusion of liberation theology and religious feminism, since they are related in both asserting that theology begins in praxis (shared action), they would be puzzled by my inclusion of fundamentalism. Is not fundamentalism simply a reactionary movement by Christian Protestants who cannot face the changes of the present and so retreat to the security of the past? The short answer is "yes" and "no." Fundamentalism in its North American clothes is both a move back in time as well as a genuinely new movement with strange and interesting new ideas.

I will conclude this section of new movements in theology by examining liberationist thinking and, in particular, its use by the feminists, beginning in the United States and spreading to Canada.

In the final chapter I will tie together some of the insights and lessons which I feel can be gained from these movements as well as point out a few of the pitfalls within each movement. In addition, some thoughts on the future impact of each on North American society will be advanced with speculations put forward about the shape of religion in the third millennium. This concluding chapter, while positioned at the end of this small book, is just the beginning of the spiritual journey on which I myself am being led by God's Spirit.

I hope that you will think of this volume as an invitation to continue to think through and to dream about your religious vision. The choices are no longer between Protestantism and Roman Catholicism, or between Christianity and other faiths. The choice that lies before us is no longer even between faith in a God and atheism. Instead, as I see it at least, the choice is between despair and hope, between a spiritual revisioning which provides new life or the slow strangulation of the spirit as the materialism and technology of the West deprive us of the spiritual oxygen necessary for life.

PART ONE

THE SCIENTIFIC TERRAIN

1

ALTERNATIVE MEDICINE

It was an embarrassing moment when the church secretary walked into my office recently. I had been experimenting with small gold-covered magnetic patches put out by a Japanese company. The theory behind the patches was that by emitting weak magnetic forces they could heal a variety of problems, ranging from arthritic pain to migraine headaches. As a migraine headache sufferer, I decided to try out the patches and had them wedged between my temples and the arms of my glasses. The secretary looked at me queerly for a moment until I realized what had caught her attention. I explained the research I was trying to do, we both laughed, and the embarrassment was dissolved.

I have reflected on the reason for my embarrassment since that incident. Would I have been concerned if my secretary had walked in on me while I was taking two aspirins? The answer is probably "no." Alternative or complementary medicine is not something I am comfortable with. In this, I am probably typical of most North Americans.

Another recent brush with alternative medical practices came about because of reading I had done on nearsightedness. Beginning as far back as the late 1800s alternative medical practitioners have claimed that it is possible to cure nearsightedness (myopia) using simple eye exercises. The novelist Aldous Huxley, author of *Brave New World*, was one prominent person who claimed to have healed his eye problems through such exercises. In the book *Healing Moments*, a collection of autobiographical stories, various women reflected on their encounters with a variety of different diseases.

One author went so far as to claim that her use of eyeglasses was a defense mechanism constructed to protect a poor self-image and was *aided* by the medical establishment who simply prescribed glasses for her myopia rather than deal with the underlying psychological causes of her eye problems. She counseled liberation from the machinations of traditional medicine through taking off your glasses and relying on Huxley's eye exercises.[1] Unfortunately, I tried this while driving and almost ended up in the ditch!

Two previous experiences with alternative medical treatments were far more positive. One used humor in the healing process. I had just finished two weeks in the hospital recuperating from the removal of my gall bladder. It was in the days when the stomach wall had to be cut open for a gall bladder operation, and I was still suffering a great deal of discomfort from the procedure. That first evening at home, however, cured my discomfort. My wife, Cathy, and I had decided to spend the evening watching the comedy *Short Circuit*, the story of a robot who took on an endearing, human personality. At one point in the movie, I began to laugh so hard that I was afraid my stitches would break loose. I lay on the floor unable to get my breath and begged my wife to turn the movie off, the pain in my side was so bad. We stopped the movie halfway through and went to bed. The next morning the pain and discomfort were totally gone. Whether the laughing had stretched the muscles back into shape, I do not know. I only know that humor worked, and I was not surprised when, a few years later, the hospital turned a room on the surgical floor into a "Humor Room" complete with a VCR and the latest comedies.

My other encounter occurred following another operation. While in the recovery room, I suffered a severe reaction to some of the medication I had been given, frightening everyone. I remember none of it except for one moment when a nurse held my hand. It was the strangest sensation for, although only dimly awake, I was aware of an energy being transferred through her hand to me. I clutched on to her for dear life. I can still recall that she had to pry her hand loose because presumably she had much to do and I was tying her down. And yet that contact was the turning point in my recovery.

As a minister, I have often experienced the same need to pry my hand loose from some parishioner I have been visiting because I had to keep another appointment. Prior to my own hospital experience, I had simply assumed that people wanted to hold my hand for comfort, for a feeling that they were not alone in their situation. Since my operation, I have begun to realize that there is an energy exchange of sorts. This explains why so many clergy find themselves fatigued after visitation, even though they have undertaken rather sedentary tasks such as simply sitting in a chair, holding someone's hand and talking to them.

It is interesting that my hospital experience would sensitize me to something which is part of the Christian tradition and which I should have been familiar with, having already finished my training and served some ten years as a Christian minister. The biblical story of the woman touching the hem of Jesus' garment is well known (Luke 8:43–48). Yet, like many seminarians, I had reinterpreted the healing stories in the Bible in light of modern-day psychiatric practices. This was a little more difficult to do in the case of the woman touching the hem of Jesus' garment, since she suffered from some sort of menstrual disorder rather than a psychiatric illness, so I conveniently ignored this story as some sort of anomaly. After my hospital stay, I began to speculate that perhaps the woman had experienced an energy transfer from Jesus to her. Jesus, more in tune with his body and himself than I am, sensed this, turned around and asked, "Who touched me?" Certainly, the disciples were confused by his question. Crowds of people were pressing around him and it was nonsensical in their view to ask who among that crowd had touched him; he was being jostled from all sides. Yet somehow Jesus knew, and somehow the woman was cured.

THE GROWTH OF ALTERNATIVE MEDICINE

Interest in alternative medicine or complementary medicine, as it is sometimes called, has been steadily growing. The reasons are varied. In part, the growing bureaucracies within the traditional medical establishment have left people feeling dehumanized, mere machines whose relationship to their doctor is often less personal than their relationship to their automobile mechanic. In his book on the growth of alternative medical practices, Tom Harpur, the journalist and theologian, relates the story of a woman who wrote to tell him about her frustrating encounter with the medical system. When she attempted to approach her doctor who was treating her for infertility he burst out, "Don't bother me with that; I'm an obstetrician and not a psychiatrist. Anything other than your fertility is not my concern." Harpur concludes:

It's precisely this cutting up of people into dozens of pieces, each belonging to some specialty or other, that has caused so many to feel cheated in their quest for healing. They may be laypersons, but they understand intuitively the truth that health has to do with total wholeness. The present outpouring of books and the rapidly increasing number of groups dedicated to holistic health is anything but accidental. They are the direct result of massive, popular frustration and dissatisfaction with the truncated and too-often dehumanized approach of conventional medical science.[2]

Another factor has been the growing emphasis on prevention rather than cure. This emphasis is not new. The old adage of medicine was "to cure, rarely; to relieve suffering, often; to comfort, always."[3] However, in recent years the emphasis on providing comfort and relief has been overwhelmed by an emphasis on curing illness. As a consequence, preventive care has taken a back seat. It has been bypassed by a medical establishment entranced with too much technology and too many drugs. In my pastoral visitation of sick people, I often hear the phrase "the cure is worse than the disease." This is particularly prevalent when individuals cope with the effects of aggressive forms of chemo- and radiation therapies in the treatment of cancer.

In reaction to such overemphasis upon a *cure*, a growing number of people have called for a return to an emphasis upon *care*. This turn toward care leads logically to a renewed emphasis upon preventive treatment. This has been greeted warmly by politicians, particularly in countries such as Canada that have a state Medicare system. As medical bills have skyrocketed, politicians and health professionals have begun to promote various schemes that would reduce the cost to society. In the province of Nova Scotia, for example, the government sees home care as a cheaper alternative to hospital care. Consequently, home care has been vigorously promoted, along with the establishment of regional and community health boards, which would be closer to the grassroots and better able to deal with health problems in a cost-effective manner. This desire to save money has also benefited alternative medical practitioners in that alternative treatments are seen to be financially less expensive than conventional medicine.

For example, the average U.S. physician-attended birth, according to the Health Insurance Association of America, cost $4,200 while the average cost of a midwife-assisted birth was $1,200.[4] In regard to back problems, Richard Knox reports in the *Boston Globe* that the Federal Government Agency for Health Care Policy and Research (AHCPR) estimates that Americans could save over $1 billion annually if only 20 percent of medical practitioners followed the agency's recommendation to use spinal manipulation for the alleviation of acute lower back pain, a procedure similar to what is already being practiced by chiropractors.[5]

Less important but still influential has been the dethronement of the medical profession from its previous position of power and respect. This trend is just beginning to affect the medical profession, which has been buttressed by its close connection with the field of science. Of all the professions, with the exception of engineering, medicine was and is most solidly grounded in the scientific world view. Consequently, it has not been until challenges to this world view have arisen within the field of science itself that the profession of medicine has come under attack.

On a practical level, this means that individuals are seeking to take control of their own health regime. They are not content any longer merely to trust their doctor to do what is right. They want to know what is wrong and to be given the information they need to make informed decisions. Many doctors have found this threatening and have retreated into medical jargon, which is one way the professions have buttressed their claim to special status and insight (for example, witness the gobbledygook language of theologies). Others have welcomed the change and have been actively promoting accountability and openness to alternative treatments.

One such individual is Larry Dossey. Born in what he calls the buckle of the Bible Belt, in the state of Texas, Dossey confesses that he was a deeply religious child who attended the local Baptist church and was involved in periodic crusade meetings until he began his university career. It was there that he was exposed to the scientific world view. Medical training followed his undergraduate work and his practice of the Christian religion disappeared altogether. It surfaced again after years of work as a medical doctor, albeit in a form more sympathetic to New Age teachings than traditional Christianity. Dossey began to experiment with the use of prayer as a healing technique. Too embarrassed to pray with his patients and reasoning that in Texas most of them were praying anyway, Dossey would begin his work day by praying for his patients before seeing them later in the day.

As part of the ritual I devised [he writes], I would shake several rattles and gourds, paraphernalia used worldwide by shamans and healers to "invoke the powers." These curious objects had been given to me by patients and friends over the years. When I used them, I felt a connection with healers of all cultures and ages. Although I never imagined that I – a white-coated, scientifically trained modern doctor – would be behaving like this, my prayer ritual was deeply satisfying.[6]

Dossey, in his writings, does not seem to have imbued the rattles and gourds with any magical powers. Instead, they function as a symbol that his return to the use of prayer was not a return to his supposedly fundamentalist Baptist background. Because that tradition would consider gourds and rattles as satanic, Dossey used them. In this, he seems typical of many people who have embraced alternative medical treatments and the religious vision undergirding them as a way of reaffirming faith without returning to the traditionalistic Protestant upbringing of their youth.

A further factor in the rise of alternative medical practices has been the inability of modern medicine to eliminate illness. It was ludicrous ever to think that would be the case. Nonetheless, the belief in inevitable progress, an integral part of the rise of the scientific world view, encouraged us to think that not only were we getting morally superior, but we were becoming physically healthier as modern medicine eliminated disease after disease. Such has not been the case. As old diseases have been dealt with, new diseases have taken their place. The AIDS epidemic has brought this fact home in a frightening and powerful fashion. The fact that the death rate from breast cancer has remained almost unchanged in spite of massive spending and the increased use of mammograms, along with radiation and chemotherapy, also has caused many to feel that something is wrong. Disease is not disappearing and although we know it not to be true, it sometimes feels that for every step forward medically, we have taken two steps back.

The effect of the failure of modern medicine to eradicate disease is aggravated by a growing inability of modern men and women to deal with the reality of death. In the past, relatives would often die at home surrounded by their family members. Even young children were exposed to death; it was part of life. For many this is no longer the case. I often ask study groups I have led if anyone has been with someone who died and only a minority responds that they have. Death is hidden away, but precisely because it is hidden away, it becomes even more frightening. No longer part of life, it becomes that dark, dirty secret that sex once was. Many people have been led to believe that new medical technology and drugs would steadily push back the frontiers of death. This belief heightens the disappointment they then feel with conventional medicine and encourages them to turn to alternative treatments.

A member of one of my churches who died from cancer typifies many who, near the end of their life, turn to alternative medicine in an effort to forestall death. His family doctor gave him permission to apply for and use an experimental medicine that was supposed to heal cancer, or at least retard its growth. The medicine did nothing of the kind but simply cost the family money. At best, it provided a few more days of hope after the medical doctors admitted failure. My own observation was that it almost robbed the dying man of the few days he had left in which he could say his goodbyes to his family. Fascinatingly, the man's family members did not feel this was true. Instead, their use of alternative medical treatments gave them some sort of inner peace. At first I speculated that this was simply due to the fact that their willingness to go the extra mile and use an alternative treatment signaled to them, and the larger community as well, that they had tried their best to help their father. As I began to study alternative medicine, however, I began to realize that the metaphysical (or as I prefer, religious) vision behind many if not all alternative medical practices provides a sense of peace and oneness which is not defeated by death.

A final factor worth mentioning as to why alternative medical practices have become so popular recently has been the explosion of the Pentecostal movement worldwide and the growth of the New Age movement within North American society. Whether such religious movements form the chicken or the egg is extremely difficult to decide. What is clear is that one of the key features of Pentecostalism is an emphasis upon physical healing through the agent of Christian faith.[7]

I myself went forward at a Pentecostal revival service in Red Deer, Alberta, when I was a teenager. I went to the side of the church and not to the front where all the action was taking place, because I was making a spiritual decision to trust God and follow God which was not the thrust of the revival service. Instead, the Pentecostal evangelists who led the service claimed that 90 percent of people's health problems were caused by one leg being shorter than the other. To correct this problem, they invited people to the front and there they prayed and pulled the leg that was shorter, commanding it to "grow in Jesus' name." Even back in the late 1970s when all this took place, it was clear to me that much of Pentecostalism was concerned with physical healing rather than spiritual healing (although the two are not unconnected).

An emphasis upon healing is also an important part of New Age teachings. In fact, often New Age devotees will claim that illnesses are deliberate attempts by the spirit within to teach the person certain lessons that they would not otherwise learn. "We can make ourselves ill and we can make ourselves well, the power is ours" typifies such thinking. So dominant is the theme of healing in the New Age movement that one scholar advances the claim that the "New Age can be looked upon as a new healing movement in American culture" rather than as a strictly religious movement.[8] This may belittle other features of the New Age movement, which are as important as this emphasis on healing. What is certain, however, is that as the New Age and Pentecostal movements have prospered there has also been a corresponding openness to and use of alternative medical practices of various sorts.

ALTERNATIVE MEDICINE
AND CONVENTIONAL MEDICINE

The list of alternative treatments is varied and long. It includes such well-known practices as prayer and anointing with oil, chiropractic manipulation of the spine, acupuncture, and homeopathic herbal medicines. These, however, are just the tip of the iceberg. Even in the small community in which I live, a variety of alternative or complementary health treatments are available, such as acupuncture, chiropractic, massage therapy, naturopathy, tai chi, herbal therapy, reflexology, electrolysis, and hypnosis.[9] In urban centers, the extent of alternative medical practices is much broader and includes aromatherapy, iridology, ayurvedic medicine, craniosacral therapy, homeopathy, yoga, rolfing, therapeutic touch, and many others.

Many of these treatments have been around for centuries, making it seem rather presumptuous to label them as "alternative" treatments. Since the adoption of the Flexner Report in the United States, though, Western medicine has been divided between conventional or mainstream medicine and alternative medicine.[10] The Flexner Report was written by Abraham Flexner, who attempted to standardize medical training in the United States. The report, published in 1910, did not deal directly with the question of what formed proper as opposed to improper medical treatment. Its mandate was to decide which medical schools in the United States should be safeguarded, and which should be closed down in order to improve American health care. Supported as it was by the Rockefeller Foundation, in cooperation with the American Medical Association, the Flexner Report effectively divided Western medicine into two competing camps. Practitioners of alternative medicine depict the impact of the Flexner Report in negative terms, while practitioners of conventional medicine claim that the

Flexner Report put medical training and medical treatment on a firm scientific basis. Whatever one's personal bias, it is a historical fact that the Flexner Report and the work of Abraham Flexner resulted in the closure of 80 percent of the then-existent medical schools.[11]

Alternative medical treatments, then, are treatments that are not taught in licensed medical schools or covered by standard medical insurance. This is changing, as various forms of alternative treatments, particularly chiropractic, are recognized and funded, although usually at a lower rate than standard treatments. Nonetheless, this simple definition is a starting point in the categorization of medical treatments. It is, however, only a starting point in that the more interesting distinction between conventional medicine and alternative medicine is philosophical and religious.

RELIGION AND
ALTERNATIVE MEDICINE

Simply put, practitioners of alternative medicine, and the patients who seek such treatment, have adopted a view of the human body as a vibrating energy field rather than as a biochemical machine. More importantly, they have adopted this view because of religious and often philosophical reasons rather than strictly scientific ones, even though an appeal to science is used to help reinforce their support of alternative medical practices.

Thus, even when this energy field is depicted in scientific terms, it is imbued with godlike characteristics. In the writings of Wilhelm Reich, for example, the language is primarily scientific but, clearly, the background thinking is religious or philosophical. Reich postulated "a mass-free energy which is omnipresent in the atmosphere, is related to the sun, fills the whole of space in varying concentrations, is drawn or used by all organisms, and accounts for the pulsing contraction and expansion of all living things."[12] In Pentecostalism, while appeal to science is secondary, the third person of the Christian trinity – the person of the Holy Spirit – functions in a similar fashion to Reich's pervasive energy.

It is the existence of this energy field (or omnipresence of the Holy Spirit) which accounts for the emphasis on oneness or holism so characteristic of modern medical treatments. At times, this seems almost painfully reductionistic, as with reflexology or iridology. It takes a great leap of faith to believe that by pushing on part of the sole of my foot, in the case of reflexology, I can control the operation of my liver. Or in the case of iridology, that the color of my iris signals the fact that I have gallstones. However, when viewed in the context of holistic thinking, such claims become more believable or at least understandable. The part affects the whole and the whole affects the part. Fix the part and the whole is also fixed. The life energy radiates throughout the body and indeed throughout the cosmos.

Perhaps the best example of this is chiropractic treatments. Instructional charts used in the teaching of chiropractic medicine show a correspondence between a portion of a human spine and some other organ within the body. Fix the alignment of the spine and you fix its corresponding organ. The same holds true with the practice known as therapeutic touch. In this system of alternative care, several key energy points are manipulated in the body in order to smooth out a person's energy field or to drain excess energy that may be causing problems.

Based in part on the Hindu notion of *chakras*, energy receptors dispersed throughout the body, a therapeutic touch practitioner will go through four steps in the treatment of their patients. First, they will gather strength and focus their own energy field through the process of meditation. Second, they determine the patient's health by moving their hands over the patient's body, being careful not to touch the patient since the energy field is seen by therapeutic touch practitioners to surround the body. Third, they then try to manipulate this energy field by massaging the air, seeking to remove blockages that may impede the flow of energy through a person. Fourth, they will try to drain off excess energy through a process known as grounding. Finally, they will check the person's energy field once again to detect whether the person is healthy or not.

The religious doctrine of the omnipresence of the divine undergirds almost all forms of alternative medicine. With refreshing honesty, Larry Dossey, the Texan doctor who advocates the use of prayer in healing, admits that, "the primary reason to focus on the role of prayer in healing is not to prove its effectiveness scientifically... The best reason goes deeper: *Prayer says something incalculably important about who we are and what our destiny may be.*"[13]

It is this religious background which forms the essential difference between conventional and alternative medicine and which lies behind the controversies between the two. Conventional medical practitioners may be religious in their personal lives but they have accepted the separation of religion from science in their medical practices. They view alternative medical practices as harmless superstitions at best and harmful quackery at worst. On the other hand, those who practice alternative medicine claim that conventional medicine has ignored the greatest healing power available and has been willfully negligent and blind.

The fact that alternative practitioners believe in spiritual forces of healing leads to the further belief that spirit interpenetrates matter. As mentioned previously, many medical practitioners and researchers are committed believers in their particular religion, but have separated the power of the divine from the power of medical technology and human ingenuity. They may believe that the divine guides and leads, but they do not believe that the divine enters life and heals. Thus, criticisms of alternative medicine which claim that nothing new is being offered and that conventional medicine also counsels good dietary practices, social involvement with others, and a positive mind set, miss the main point. The point is that after years of separation, the fields of religion and medicine are being reunited in alternative medicine.

Interestingly, this marriage of modern medicine and ancient religion is not confined to one form of religious expression. The New Age movement and Pentecostal forms of Christianity could not be more antagonistic to one another, yet both forms of religion are influential in various alternative medical practices. At the risk of being too simplistic, the four main religious strands which surface most often in alternative medicine are Pentecostal Protestantism and its charismatic Roman Catholic counterpart, theosophy, New Age movements, and various elements of Eastern religious traditions.

This eclectic stewpot of religious faiths is viewed with deep suspicion by orthodox believers as well, but for very different reasons than those who hold to the separation of religion from science. The former feel that the adoption of beliefs from other religious traditions signifies degeneration and apostasy, that is, a forsaking of the one true faith for counterfeit beliefs. The latter feel that the separation of religion from science is one of the hallmarks of Western civilization. Many within this second group seem to have no religious faith whatsoever. They are happy to keep the two fields of religion and science separate so as not to dilute science which functions according to the rules of logic and reason, with religion which is seen as an ethereal fancy with no rational grounding. However, in spite of the distinct differences between these two groups they share a common antagonism toward alternative medical practices which they label as silly at best and dangerous at worst.

WHY ALTERNATIVE MEDICINE IS SO INTRIGUING

It is precisely the eclectic nature of the beliefs behind alternative medicine which make it an interesting topic of study for me. For some time, I have been convinced that traditional Christianity as I experience it does not speak to the deep longings of the typical North American. For a long time I accepted the view, prevalent within the conservative Protestant tradition of which I was a part, that this was a sign of the apostasy predicted in the New Testament as a prefiguration of the end of the world. This negative attitude meant that I could safely write off not only alternative medical treatments, but also the belief structure behind those treatments as unworthy of study or reflection.

One of the key factors in changing this negative position has been a fragmentation that I have felt within myself. The specialization characteristic of the modern university which has been so much a part of my life, not to speak of society in general, leaves one feeling as segmented as an orange, held together only by the mere rind of one's will. There is no center. Faith does not relate to science. Politics is divorced from ethics. Economics is reduced to financial manipulation. The average churchgoer inhabits one world on Sunday morning and another during the workweek. One's work and one's family life seem to have no connection to each other.

Can unity be found in the midst of diversity? Alternative medicine is attractive precisely because it seeks to unite mind, body, and soul, claiming that spiritual and physical health affect each other and that harmony is the key to health. Granted that many of the treatments and the beliefs behind alternative medicine seem simplistic, and that claims to the interaction between modern physics and Eastern religion are overdrawn. Nonetheless, deep speaks to deep; the soul hears her name being called and she responds.

Alternative medicine is worth studying because there *does* seem to be some sort of energy or spirit beyond the ability of modern science to detect – although this energy has always been a part of the belief structures of the major religious traditions. Conventional medicine will tell you that when two people fall in love with each other it is merely a chemical reaction within each signaled to the other by subconscious and almost undetectable gestures and signs. This explanation is unsatisfactory to the person who falls in love and who knows that more is going on than chemical reactions. It begins to break down completely when one considers the energy present within a group of people.

As a minister, I confront this every week. There are times when this energy (in theological terms the Holy Spirit) is present in the worship service and times when it is absent. Like the wind, the Spirit blows when and where it will. I do not understand it, but I can see its impact much as I can see the leaves of a tree rustle with the passing breeze. There is something present when a group of a people gathers that is greater than the sum of the parts. As Jesus put it, "where two or three are gathered in my name, I am there among them" (Matthew 18:20 NRSV).

It is both frustrating and comforting that I cannot control the presence or the effect of this energy. Frustrating, in that when the Spirit is present the sense of unity and joy is such a wonderful feeling, that one would like to maintain that state forever; much as Peter, James, and John wanted when they witnessed the transfiguration of Jesus and burst forth with the suggestion that Jesus and they build three shelters on the mountaintop and live there forever. It is comforting, because if I could control the presence and activity of the Spirit, then this energy would simply be another physical phenomenon, exciting in and of itself, but basically without meaning in the light of the larger questions of life and death.

ALTERNATIVE MEDICINE
AND THE LARGER QUESTIONS OF LIFE

Some practitioners of alternative medicine would be content if this energy proved to be simply an undiscovered fifth force which complemented the previously known four forces of nature. Then, beside gravity, electromagnetism, and weak and strong nuclear forces, scientists could simply add a fifth force of nature. However, most practitioners of alternative medicine, I suspect, would not be happy if this proved to be true. The holistic thrust within alternative medicine pushes beyond the boundaries of death to what have traditionally been religious issues – questions of meaning and purpose – ultimate questions.

It is this sense of being connected with something greater than you are, something cosmic, (in religious language, God) which accounts for much of the appeal of alternative medicine. You – your destiny, your health, your future – count. Death cannot extinguish your life; it can only alter the contours of the energy that forms who you are. Often this appeal to questions of ultimate importance is hidden within alternative medicine.

An example of this occurred in the city of Halifax recently. A popular newspaper columnist went to a Benny Hinn healing crusade. Benny Hinn, typical of many within the Pentecostal tradition, travels the circuit in North America, holding healing services and selling his books and audiotapes. Rather than offer the typical dismissal of this healing crusade, the columnist, Peter Duffy, was quite open to the possibility of healing occurring at the crusades. This resulted in several critical letters, most pointing out the manipulation and financial skullduggery associated with such crusades.

Others objected on religious grounds, noting that in the Christian tradition, it was Jesus who performed healing and since Jesus was no longer on earth, such miracles had ceased.

In a subsequent column, Peter Duffy explained that he did not support the type of behavior typical of evangelists such as Benny Hinn, but, nonetheless, he believed that sometimes such crusades awakened us to our own inner power to affect self-healing. On the surface, Duffy, like many others, has confined alternative medical practices to the same level as conventional medicine, and yet the example he provides as to why he was open to the Benny Hinn crusade, in the first instance, is revealing. His father-in-law was diagnosed with cancer and not given very long to live. Some six months prior to his death, his father-in-law attended a healing crusade. In light of his father-in-law's death a few months later, one would have suspected this event to have soured Peter Duffy on faith healers. It did nothing of the sort. He writes:

One night he went to a healing crusade and came home euphoric. He told us that he was able to run around at the back of the auditorium without his oxygen.

He died six months later, but I've never forgotten how this man we loved found hope, however briefly. And whenever I'm tempted to scoff at those who claim to heal in the name of Jesus, I remember my father-in-law and pause.[14]

Many alternative medical treatments focus on chronic pain and the management of such pain. It is in confronting life-threatening illnesses, however, that the religious substratum beneath alternative medicine is most clearly visible. Even when alternative treatments fail, they provide relief and hope. This logical contradiction is only possible because they are based on religious beliefs which deny that death is the end of life as we know it and also on the belief that illness is not the enemy of health but often its friend, in that it signals to the person that their energy is out of alignment. The faith healer may only be a lightning rod to our own inner power, as Peter Duffy puts it, but that inner power connects us with the divine and assures us that death is not the end but merely a transformation.

It is this religious belief in life after death (whether the typical Christian belief in the resurrection of the dead, or the Eastern belief in reincarnation and the ultimate unity of all things) which explains the popularity of writers such as Bernie Siegel. Basically, Siegel's philosophy is a mixture of positive thinking, self-affirmation, and personal control over one's own health. On the surface, religious beliefs in the nonfinality of death are absent or downplayed. And yet, the sustained popularity of Siegel's books in light of scientific data, which reveals no correlation between taking control and prolonging life in the case of cancer,[15] testifies to the presence of religious hope.

I can find no other reason why so many people have turned to alternative medical practices. According to the January 28,1993, edition of the *New England Journal of Medicine,* 34 percent of Americans used one or more forms of alternative treatment in 1990. Even more surprising is that the more educated were the ones most likely to seek out alternative care! As John Langone reports in a special issue of *Time* magazine, this means that in the United States alone around $13.7 billion were spent on "a bewildering array of breakaway treatments, including chiropractic, colonic irrigation, meditation, homeopathy, naturopathy, hypnotherapy, music therapy, folk medicine, guided imagery and shiatsu massage."[16] In Canada, studies show that 20 percent of Canadians use some type of alternative care.[17]

The obvious rebuttal is that such a high percentage of people turn to alternative treatments because they work. Yet, while anecdotal evidence abounds, scientific evidence is scattered and still in its infancy. As Philip and Mary Jo Clark put it, "in the final analysis, the current research base supporting continued nursing practice of therapeutic touch is, at best, weak. Well-designed, double-blind studies have thus far shown transient results (Grad, 1961), no significant results (Randolph, 1980), or are in need of independent replication (Grad, 1963)."[18] I suppose that there *is* scientific evidence at all is suggestive. Moreover, the work done by McGill University researcher Bernard Grad in collaboration with a faith healer named Oskar Estebany is impressive.[19] Still the number of such studies is minuscule, with most of them appearing in parapsychology journals operating on the fringe of the academic and scientific establishment.

What is happening with alternative medicine, then, is that it is becoming the door by which people are able to reaffirm latent religious beliefs that were repressed and ignored. Suddenly, there does seem to be some proof that the divine exists within us, and if the divine exists within us then it makes sense that the divine exists outside of us as well. We are not alone any longer in a cold, meaningless universe where we live a few years and then pass into oblivion. Our life has meaning and purpose and eternal significance. As religious studies professor, Robert Fuller, puts it, "the turn to alternative medicine is thus in large part a turn to alternative metaphysics."[20]

SOME CONCLUSIONS

The impact that alternative medicine will have on conventional medical practices is as yet unknown. Moreover, it is beyond the scope of my expertise and this book. The impact that alternative medicine will have on religious practices and beliefs is also uncertain. However, here I am on more familiar terrain and thus willing to make some predictions.

The first is that a holistic emphasis on the interconnectivity between the human mind, body, soul, and the divine will continue to grow. This will cause enormous conflicts in our North American society, which is based on the separation of church and state. Since the state regulates medical treatment in both the United States and Canada, alternative medicine will force a reexamination of how the separation of church and state can continue. When writers such as Larry Dossey claim that for doctors not to use prayer in their medical practices will constitute medical malpractice, it becomes clear that the rules of the game are being challenged.[21]

To some extent, the pluralistic background of alternative medical practices will lessen this conflict. Since alternative medical practices claim inspiration from various religious traditions, both Western and Eastern alike, the battle over the entrenchment of the Christian faith will no longer be the focus in the future. However, the aggressive secularity of North American society in many elements of government, media, and academia is not happy with any talk of the reality and importance of faith, be it Western or Eastern. The conflict will change its focus, but not its intensity.

Moreover, this holistic emphasis on the interconnectivity of the mind, body, spirit, and the divine will spill over into other areas. If the individual cannot be fragmented then it stands to reason that society cannot be fragmented either. Political boundaries will continue to decrease in importance. The environmental awareness of a symbiotic relationship between the human creature and the natural world will remain prominent and grow in popularity. The emphasis on specialization in both the academic world and the business world will shift to an emphasis on generalization. The biblical definition of wisdom as right living rather than as an accumulation of facts will again be in vogue.

In the field of religion, pluralism and eclecticism will grow, bedeviling the doctrinal purist. People will hold logically incompatible doctrines without feeling any intellectual discomfort. The Pentecostal emphasis upon the Holy Spirit will become dominant within the Christian tradition. Denominational groups which cannot find doctrinal room for the activity of the Spirit will be marginalized and in some cases disappear. The Age of the Spirit and the Age of the Father will displace the Age of the Son.[22] An emphasis on sin and redemption will give way to an emphasis upon creation and recreation.

These changes can and are already happening regardless of whether or not there is a benign and good cosmic energy, a divine spirit, which permeates the world. Once people's world views begin to shift, old visions begin to die away or are forced to share space with new visions. If, however, this divine energy does exist and can be manipulated by the human creature, even more dramatic changes can be envisioned. In the New Testament, Jesus tells his disciples that if they ask a mountain to be moved from one place to another it will happen. Most theologians and preachers, myself included, would reinterpret this passage, and point to the fact that if we work together as human beings we can and *have* moved mountains. But if human beings can harness the divine energy, which alternative medicine believes in, then this circumlocution is no longer necessary. The text stands

as Jesus said it: "have faith in God. Truly I tell you, if you say to this mountain, 'Be taken up and thrown into the sea,' and if you do not doubt in your
heart,... it will be done for you" (Mark 11:23 NRSV).

This, of course, seems far-fetched, like the tales found in tabloid news-
papers. Immediate objections to the manipulation of the natural world come
to mind. If I can move a mountain, so can you. What if you want this
mountain here and I want it there? Moreover, what happens to the inhabitants of that mountain or to the surrounding environment or to the sea where
it is dumped? Without a stable, common field, the world becomes a chaotic
and frightening place. Yet this is precisely the type of world which scientists
claim exists on the quantum level. It is not illogical to suppose, then, that
the exercise of this energy can coexist with the world as we know it in much
the same way as quantum physics coexists with Newtonian physics.

Before we examine the world of quantum physics, however, we will
look at another quasi-scientific movement that is growing in importance in
our society. The Gaia theory, in which the Earth is seen as a living being,
started with humble and focused beginnings, a scientific theory and nothing more. It has not stayed that way as environmentalists concerned about
the future health of the Earth and feminists seeking a role for the Goddess
have been attracted to it.

2

GAIA –
THE GODDESS EARTH

It is interesting to speculate what would have happened to its popularity if James Lovelock's theory that the Earth is a living organism had been called the "Biocybernetic Universal System Tendency/Homeostastis" as he had originally intended, instead of the "Gaia Theory," the term by which it became known.[1] A neighboring villager, the novelist William Goulding, suggested the name Gaia to James Lovelock, a peripatetic scientist. Lovelock agreed and the theory of the earth as a living being was introduced under the name "Gaia," with all the mythological and religious symbolism that such a term carries.[2]

Ge, the goddess of the earth, and Ouranos, the god of sky, are among the oldest deities in Greek mythology. Ge was depicted as a woman, half of whose body extended from the Earth. As she received seed and rain, she became the mother from whom all life proceeds. In time, the Greek word Ge simply came to mean either the Earth as a whole, or land as contrasted with water. The word survives in the English language as the first part of the construct "geology," the study of Ge (the earth); "geography," the writing of Ge; and "geometry," the measuring of Ge.

Initially, Lovelock seems to have paid little attention to the mythological and religious symbolism behind the word his friend had bequeathed him, but others did. Although introduced as a scientific concept, Lovelock's theory has gained more support and favor in philosophical and religious circles than it has in scientific ones. However, the scientific basis of Gaia theory must not be discounted too quickly, as we will examine shortly.

Almost prophetically, in light of the environmentalist's use of the Gaian theory, Lovelock's reputation was first established by his participation with A. J. P. Martin in developing the chemical analytical technique of gas chromatography. Lovelock added some embellishments that extended the range of Martin's invention, most notably the electron capture detector able to detect minute traces of certain chemical substances. The sensitivity of this detector resulted in the discovery that pesticide residues "were present in all creatures of the Earth, from penguins in Antarctica to the milk of nursing mothers in the USA."[3] It was this discovery which lay behind Rachel Carson's popular book *Silent Spring,* an important book in launching the environmental movement in North America.

Lovelock traces the genesis of the Gaia theory, however, not to his work with Martin but to his work with the Jet Propulsion Laboratories of the California Institute of Technology in the 1960s. Funded by a National Aeronautics and Space Administration (NASA) program, Lovelock was part of a team whose task it was to determine whether there was life on the planet Mars. When he and colleague Dian Hitchcock were assigned the task of critically examining the life-detection experiments proposed for Mars, Lovelock grew frustrated with the project. "It seemed," he wrote, "as if the experiments had all been designed to seek the sort of life each investigator was familiar with in his own laboratory. They were seeking Earth-type life on a planet not in the least like the Earth. To Dian and me, it seemed that we were guests on an expedition to seek camels on the Greenland icecap or of one to gather the fish that swam among the sand dunes of the Sahara." In response, Lovelock and Hitchcock set out to devise a more general form of life detection which in Lovelock's words, "would recognize life, whatever its form might be."[4]

Lovelock and Hitchcock came up with the idea of atmospheric analysis as a way of detecting the presence of life in a manner that would be holistic and broad ranging. When this method was applied to the earth's atmosphere, Hitchcock and Lovelock became convinced that the "only feasible explanation of the Earth's highly improbable atmosphere was that it was being manipulated on a day-to-day basis from the surface, and that the manipulator was life itself."[5] The problem was that there was no known life force capable of regulating the atmosphere to such a degree. Therefore the Earth had to be doing this for itself, and, according to Hitchcock and Lovelock's model, the Earth had to be a living being; or as another Gaian author, Elisabet Sahtouris puts it, the Earth had to be thought of as a "live planet rather than a planet with life on it."[6]

This conclusion – that the world was a living being – did not come to Lovelock all at once. In 1965, the Martian exploration program was suspended and thus attempts to find life on Mars were discouraged. However, Lovelock's work in atmospheric analysis prompted the research branch of Shell Oil to invite him the following year to consider the possible consequences of air pollution from such causes as the ever-increasing rate of combustion of fossil fuels.[7] Freed to turn his attention from an analysis of Mars to that of the planet Earth, Lovelock made his dramatic conclusion. He writes:

Working in a new intellectual environment, I was able to forget Mars and to concentrate on the Earth and the nature of its atmosphere. The result of this more single-minded approach was the development of the hypothesis that the entire range of living matter on Earth, from whales to viruses, and from oaks to algae, could be regarded as constituting a single living entity, capable of manipulating the Earth's atmosphere to suit its overall needs and endowed with faculties and powers far beyond those of its constituent parts.[8]

This hypothesis began to take on the status of a theory. In 1969, Lovelock presented a paper at a scientific conference in Princeton, New Jersey. By Lovelock's own admission, the Gaia hypothesis did not have much of an impact. However, one person heard and responded to Lovelock's hypothesis, filling in a vital deficiency in Lovelock's analysis. In contrast to Darwin's "survival of the fittest," Lovelock's hypothesis necessitated greater cooperation as the various elements of the Earth, plant, mineral, vegetable, and human life cooperated to ensure the stability of the Earth as a living entity.

Lynn Margulis, a scientist from Boston University, provided this missing link. Margulis had become convinced in her studies that cooperation rather than competition was the key for understanding evolution. In 1973, Lovelock and Margulis co-authored two papers postulating that the climate and chemical composition of the Earth's surface environment is, and has been, actively regulated at a state tolerable for the *biota*.[9] In 1979, Lovelock wrote a book simply entitled *Gaia*, published by the Oxford University Press, and the Gaia theory was born.

Since the publication of the two articles co-authored with Margulis, along with his 1979 book, Lovelock has been busy on two fronts. In a fashion analogous to feminist and liberation theologies discussed later in this book, Lovelock, and a growing number of Gaian advocates, began to construct a history of the Gaian concept. James Hutton, a British geologist whose book *A Theory of the Earth* (1795) formed the basis of modern geology, is cited by Lovelock as an early scientist comfortable with the "notion of a living Earth."[10] Another thinker, a Russian philosopher Y. M. Korolenko, is also listed as an important historical figure in the Gaia movement mainly as a result of his influence on his nephew, the Russian scientist Vladimir Ivanovitch Vernadsky, who influenced the biologist G. E. Hutchinson.[11]

Besides constructing a history of the idea, Lovelock has also been busy seeking to distance himself from a host of academics, writers, and religious figures who have adopted his theory for other than scientific reasons. William Irwin Thompson refers to such people as the "gooey Gaians," noting that it is unfair "to blame the Goddess cult of the New Agers at Findhorn or the Cathedral of St. John the Divine on Jim [Lovelock] and Lynn [Margulis]."[12] Lovelock himself, has admitted that his initial language was not precise enough.

I have frequently used the word Gaia as shorthand for the hypothesis itself, namely that the biosphere is a self-regulating entity with the capacity to keep our planet healthy by controlling the chemical and physical environment. Occasionally it has been difficult, without excessive circumlocution, to avoid talking of Gaia as if she were known to be sentient. This is meant no more seriously than is the appella-

tion 'she' when given to a ship by those who sail her, as a recognition that even pieces of wood and metal when specifically designed and assembled may achieve a composite identity with its own characteristic signature, as distinct from being the mere sum of its parts.[13]

In spite of Lovelock's protests, though, he has been happy to be identified with the Cathedral of St. John the Divine. At the invitation of its rector, James Morton, Lovelock preached there on the occasion of the presentation of Paul Winter's musical piece entitled the *Missa Gaia*. Moreover, without the support which the "gooey Gaians" have given to Lovelock's idea it would never have become the topic of popular discussion that it now is.

GAIAN THEORY AND TRADITIONAL SCIENCE

The debate over whether the Gaia theory is a serious scientific theory is a heated one. The 1988 meeting of the American Geophysical Union, an international gathering of geologists and geochemists, was devoted completely to a discussion of Lovelock's and Margulis's views. At that meeting, the Gaia theory received both "vigorous criticism" and "vigorous support."[14]

This criticism arose not only due to the activities and support offered by those more interested in the religious rather than scientific implications of the Gaia hypothesis, but also from Lovelock himself. Although he has stated categorically that Gaia is not to be thought of as a sentient, self-reflecting being, the language he uses is more often poetic than it is scientific. Often Gaian theory seems to be driven by philosophical concerns as much as scientific ones. Lovelock is frank concerning his biases. In his 1979 book he confesses,

The Gaia hypothesis is for those who like to walk or simply stand and stare, and to speculate about the consequences of our presence here. It is an alternative to that pessimistic view which sees nature as a primitive force to be subdued and conquered. It is also an alternative to that equally depressing picture of our planet as a demented spaceship, forever travelling, driverless and purposeless, around an inner circle of the sun.[15]

Moreover, the populist style in which the Gaia hypothesis is presented also lends support to those who would deny that it is a valid scientific theory. A common image in Gaian literature, coming from the world of the arts rather than the scientific world, asks the reader to imagine a film of the Earth shown at high speed. This film always begins with a serene pastoral setting and ends with belching smokestacks and urban sprawl. While an effective method, the scientific underpinnings seem somewhat questionable.

In defense of the scientific nature of the Gaia hypothesis are those who admit that while the language seems somewhat unscientific, the reason is not because the Gaia hypothesis is pseudoscience, but because it is a new approach in scientific investigation; one that deliberately directs its attention to the whole planet as the final destination of scientific inquiry. This new approach, though, only confuses the issue in that the hypothesis cannot be tested as it presently stands. For some, this is proof that the theory is not a scientific one. The inability of scientists to prove various elements of the Gaia hypothesis, however, is not the important issue. After all, many of the hypotheses within the field of quantum physics cannot presently be proved either. More important than its provability by the senses, the Gaia hypothesis directly contradicts two key scientific theories. The Gaia hypothesis presents a challenge to Darwin's theory of evolution as well as to the second law of the theory of thermodynamics. This challenge is of far greater significance in determining whether Lovelock's hypothesis is scientific.

The attack on Darwin's theory of evolution arose out of the work of Lynn Margulis on cell structures and activities. According to Darwin, competition propels the evolution of animal species (including the human creature). According to Margulis and Lovelock, cooperation rather than competition is the key to understanding the ongoing evolution of life. As Margulis asserts, "we consider naïve the early Darwinian view of 'nature red in tooth and claw.' Now we see ourselves as products of cellular interaction... Partnerships between cells once foreign and even enemies to each other are at the very roots of our being."[16] Advocates of Gaian thought claim that Darwin's theory of evolution was based in large part on the social conditions of his day. A symbiotic relationship, they criticize, existed between Darwinian theory and Victorian England. Social Darwinism, in which the prosperity of the rich was defended as the engine that fueled economic and societal growth, was not just a by-product of Darwin's theory of evolution, but a cooperating partner. In our present day, Gaians add, as we have been able to unmask the ideological motives behind past scientific theories, we have come to see that nature cooperates as much or more than it competes. More accurately, Gaians claim that the isolated instances of competition that we observe need to be placed in a larger scheme of cooperation for the ongoing maintenance and growth of life.

Another difference with traditional scientific thought is the challenge posed to the second law of the theory of thermodynamics. According to this law, in a closed system, entropy (disorder at the microscopic level) always increases; this provides time with a direction. A log burns; as it burns, it gives off heat with the result being that the ordered energy within the log becomes the disordered energy of heat. More poetically, T. S. Eliot writes in *The Hollow Men*, "this is the way the world ends, this is the way the world ends, this is the way the world ends not with a bang but a whimper."[17]

Gaian advocates must dispense with this theory, however, since it leads to the inevitable end of the world by heat death and gives the lie to the belief that Gaia is a self-regulating mechanism that can continue to regulate its life forever. As Theodore Roszak concludes "'time's arrow,' once thought to be exclusively oriented toward the heat death of the universe, points in the direction of structure and systems, and at last conscious life: the human frontier. *Entropy is exhausted matter's arrow, not the living mind's.*" [emphasis mine][18] In other words, entropy governs only material things, according to Gaians, but life itself is not governed by this theory which should be confined to where it began, the study of the operation of steam engines.

GAIAN THEORY AND ECOLOGY

The challenges which the Gaian hypothesis or theory presents to science have not yet been resolved. The burden of proof, in my opinion, rests on those who would dispute the second law of thermodynamics and Darwin's evolutionary theory. Thus, at present, one would be forced to conclude that

Gaian thought is not scientific. This, though, has not stopped many people from treating it as science, particularly in the field of ecology. In part, this partnership with the ecological movement was a natural outgrowth of the relationship between Lovelock's modifications to Martin's chemical analytical technique of gas chromatography and the work of Rachel Carson. Just as important was the bridging role played by religious feminists who were attracted to environmental concerns on the one hand and the concept of the Earth as a Mother Goddess on the other.

Originally, environmentalists advocated the preservation of natural supplies and animal species in order to avoid resource depletion and to fight pollution. Feminists cast aspersions upon such motives. Rather than seeking to preserve the environment in order to ensure that we could not run out of resources, religious feminists claimed that we should care for the environment because of the moral imperative of nurturing. Traditional environmentalists were treated as dupes of the North American industrial machine who preserved simply in order to destroy. Religious feminists sided with "deep" rather than "shallow" ecology.

Miller defines these terms, noting that,

Shallow ecology is establishment ecology – the ecology of government, industry, and university – limiting its concern at the very most to the fight against pollution and resource depletion. Its central objective is the maintenance of the status quo, attempting to do a bit better environmental biology and working for the health and the affluence of people at home. Deep ecology, on the contrary, moves beyond formal reductionist biology and concerns itself with whole-system values and with the well-being of people in poorer and developing countries.[19]

It is easy to see how the Gaian emphasis on interdependency and on the whole rather than on the part appealed to feminists who had already mounted their own critique of the various ideological dualisms which divide rather than unite. Moreover, the emphasis on nature finds a receptive audience among feminists who feel that the stress on the soul and the mind, which has been the dominant stress within Western civilization, has contributed to the diminution of women.

Most religious feminists part company with the deep ecologists in regard to the view, taken from Gaian theory, that humanity is a cancerous growth on the face of the Earth and should be removed for the health of Gaia. The Gaian theory, itself, is marvelously unconcerned about this issue. At heart it is an optimistic view which believes that the Earth will control the atmospheric conditions in order to ensure its ongoing existence. If this means that humanity disappears, so be it, but the important point is that the human creature has no real influence over the activities of Gaia. The Gaian theory does not sound the call for ecological revolution, but rather for accommodation to the rhythms and movements of Gaia. According to such a view, Gaia is not threatened – rather, the human creature is threatened.

GAIAN THEORY AND THE NEW PHYSICS

The other connection often made between Gaian thought and science is with the theories and speculations concerning the new physics, examined in the next chapter. Thus, Theodore Roszak notes that the "anthropic principle" (that is, the concept that the laws of physics have a particular mathematical form which encourages the emergence of thinking biological organisms such as men and women) within the new physics has opened scientists up to an awareness and amazement concerning the combination of factors that lie at the basis of physical reality.[20] Meanwhile, Rosemary Radford Ruether, a Christian feminist and Gaian advocate, claims that the new physics at both the macro level and the micro level has broken down the previous separation of science and faith, thus allowing Gaian theory to arise.[21] More cautiously, Alan Miller asserts that because of the uncertainty principle our knowledge is always imperfect. He continues; "The lessons for biology and its related ethical systems are clear: to treat the Earth and its inhabitants carefully and with humility, treading gently indeed wherever cautious human beings should go."[22] Elisabet Sahtouris mentions the "bootstrap theory" of the new physics, claiming that it supports the Gaian concept of the interdependency of all things.[23]

Tellingly, at least to those who dispute the scientific validity of the Gaian theory, these representatives of Gaian thought do not focus on the core scientific insights contained within the new physics, but on its philosophical implications as well as on speculative theories such as the bootstrap theory, which has now gone out of favor. Of course, it is possible that the authors that I have mentioned belong to the category of "gooey Gaians," whose thought and activity should not be blamed on James Lovelock. However, in the case of Sahtouris, Lovelock wrote a foreword to her book in which he states, "Elisabet Sahtouris's conception integrates scientific Gaian evolution with the human search to connect with our roots... *she comfortably integrates the traditionally separated domains of biology, geology, and atmospheric science* [emphasis mine] to show us the evolution of our living planet and our own roots within it."[24]

Although connections exist between Gaian theory and ecological thought as well as between some of the insights of the new physics, until Darwin's theory of evolution and the second law of the theory of thermodynamics are radically changed or discarded, it is doubtful whether the Gaian theory will garner much support within the scientific community. Most scientists will continue to dismiss it, as the physicist/philosopher James Kirchner does, as "an awkward, amorphous grouping of ideas with dubious provability."[25]

GAIAN THEORY AND RELIGION

In the field of religion, on the other hand, Gaian thought is often warmly received and becoming increasingly influential, although even here it is not without its critics. Advocates of conservative Christianity view it with deep suspicion while most religious feminists and those who admire aspects of Eastern religions are supportive. A recent youth play performed in the nave of Lincoln Cathedral in England is typical of the warm reception and the influence that Gaian thinking has had within Christianity. The play is set in a postnuclear world in which food and water are scarce. The stage is strewn with litter and the characters wear facemasks in order to protect themselves from the poisonous pollution. One of the characters, "a messiah-like figure dressed in brilliant white, preaches that 'ancestors' living in the landscape must be reawakened in order to make the ground once more fertile."[26]

Key to the acceptance of Gaian thought is the criticism that traditional Christianity has encouraged the domination and subjugation of the earth and is, therefore, culpable for the current environmental crisis. The logical inference of this argument is that at its most fundamental level, the environmental crisis is a spiritual crisis that can only be solved by a spiritual solution that treats the earth as more than an empty vessel. This criticism of Christianity is not new. In his book *Design In Nature*, Ian McHarg criticizes the anthropocentric perspectives of the West and claims that Christianity,

...in its insistence upon domination and subjugation of nature, encourages the most exploitative and destructive instincts in man rather than those that are deferential and creative. Indeed, if one seeks license for those who would increase radioactivity, create canals and harbors with atomic bombs, carry poisons without constraint, or give consent to the bulldozer mentality, there could be no better injunction than this biblical creation text [Genesis 1:28]. Here can be found the injunction to conquer nature – the enemy, the threat to Jehovah.[27]

Such a criticism is nonsense, as even a cursory reading of the first creation story found in Genesis chapter 1 will confirm, even more so when the second creation story, found in chapters 2 and 3, is also examined. The intimate connection between the created order and the human creature is revealed by the fact that sin disrupts this once idyllic unity. The serpent, at one time the friend of man and woman, strikes at the woman's heel and she seeks to stomp on the serpent's head. Hard toil and labor replace the easy living in the Garden of Eden in order to provide enough food to eat. Whatever dominion means in Genesis 1:28, it does not mean exploitation but rather a mutuality.

Indeed, throughout the Bible, whenever the prophets look toward the Day of the Lord they picture it as a time when the rift between the human creature and nature is restored to its initial harmony. Isaiah, for instance, writes about the wolf lying down with the lamb and the lion dwelling with the calf and a little child leading them (Isaiah 11:6). In the New Testament, this image is taken up by the missionary-theologian Paul as he writes about the creation groaning in anticipation for its redemption, which he depicts as being contingent upon the full redemption of man and woman (Romans 8:18–21).

The exaltation of the human creature to the position of prominence with its consequent disdain for nature is a secular rather than Christian theme. It was the humanists of the Renaissance who asserted that "man is the measure of all things," and who deliberately tried to discard the sense of "unworthiness" which Christian humility seemed to require, not the Christian thinkers.

On the other hand, Ted Connor's disagreement with Gaian theory in the influential conservative Protestant journal *Christianity Today* is in its own way just as misguided as are the criticisms which lay the blame for the present environmental crisis at the feet of Christian theology. In his article "Is the Earth Alive?" he concludes that the conservative Christian must resist Gaian thought because it displaces the centrality of man with the centrality of the Earth. In the Judeo-Christian story, however, wo/man is never central, that place always belongs to God. Wo/man is part of the created order, charged to care for creation as a good steward, and always aware, as has been mentioned earlier, that her/his redemption results in and depends upon the restoration of the entire created order.

A more accurate attack on traditional Christianity as a result of the insights of the Gaia theory than that mounted by McHarg focuses on the dominance of rational, linear thinking within Christian theology and the consequent neglect of the mystical and the dramatic. An important difference between the Western tradition that dominated in Christianity and the Eastern tradition that all but disappeared concerns the flow of history. In the Western, Latin tradition, history is linear and progressive. It moves from "a" to "b," from beginning to end. The fierce arguments in North American Protestantism over whether the flow of history moves upward to new heights, or downward to Armageddon, masks the agreement that exists between the liberal and the fundamentalist. Both view history as nonrepetitive; both view history as the locus of salvation. Moses is a historical figure. The entry into the Promised Land is a historical event. Jesus is a historical figure. His life and death are historical events.

In comparison, Eastern religions adopt a cyclical view of nature in which salvation is accomplished not by involvement in historical events, but by the transcendence of history through meditation and spiritual exercises. This Eastern influence is not absent from Christian thought or from the biblical story, but it is certainly muted. Gaian theory, tied as it is to the rhythms of the Earth, is much more attuned with cyclical rather than linear thinking. Salvation is not defined historically, but rather mystically. Salvation is oneness with Gaia, with the Earth. It is peaceful surrender to that great circle of life that rolls on throughout time, transcending time and making time irrelevant.

Many Gaians are fond of using this theme of life as an ever-flowing drama. Strangely, they most often use the image of the movie camera, a rather advanced technological invention, in order to depict the Earth blossoming like a flower which slowly blackens from pollution caused by the human creature. Even in traditional Western Christianity, though, this emphasis on the dramatic has survived. In Roman Catholicism, for instance, the mass is the central act of worship; it is a dramatic act, the playing out of death and resurrection. As William Thompson, a friend of Lovelock's and an avid promoter of his thought, points out, "When Jesus takes bread and wine and says: 'Take this in remembrance of me, for this my body and blood,' he is not the masochistic psychopath that Freud made him out to be, but a poet with an ecological vision of life who is using myth and symbol to express how all life is food to one another."[28]

In this regard it should be noted that although he was condemned as a heretic, the Alexandrian theologian Origen was an important figure in his day and has become recognized recently, along with Augustine, as the most important theologian of the early church. Born and raised in Alexandria on the coast of Egypt, home of the famous Alexandrian Library, Origen rejected much of the Hellenistic and Eastern influences he was exposed to, but he also incorporated many of them into a picture of the Christian religion which is more dramatic and cyclical than the Tertullian-Augustinian tradition which prevailed. According to Origen, the salvation of the one depended upon the salvation of the totality. In his famous image, everyone would be saved, even Satan himself. From the oneness of God, we had fallen as our ardor for God cooled, but Christ like a giant magnetic force would draw us back to that oneness once again. The difference between a thinker like Origen and contemporary Gaians is that Origen felt that this process once complete did have an end. Oneness, once restored did not divide again. As well, Origen was too influenced by Platonic thought to value the natural world. His was the world of the spirit where the body was discarded as at best a hindrance and at worst the locus of sin.

The problem for both Origen and the Gaians concerns the preservation of individual consciousness. Gaians seem not to worry about this; for most of them it would be dismissed as an anthropocentric as opposed to a geocentric concern. Immortality in Gaian thought does not involve the ongoing consciousness of the individual, but rather his or her unconscious participation in the circle of life. In fact, the Christian concept of the resurrection of the individual comes under heavy attack by many Gaians. Yet I, for one, do not find my ongoing participation in the circle of life (basically the food chain of life), much consolation. Although I understand and long for the mystical sense of participation in the oneness (whether spiritually conceived as in Origen's thought, or materialistically conceived as in Gaian thought) I also want some sense of ongoing individual awareness, as, I suspect, do most others.

GAIAN THEORY AND RELIGIOUS FEMINISM

Another area where Gaian criticism of traditional Christian thought is more accurate than is the case with the fingering of Christianity as the environmental bogeyman concerns the role of the feminine. Again, a misreading, or at best a superficial reading, of the biblical text is often at fault rather than any real misogynism. However, there is a strong streak of patriarchal language and thought in the Bible. The partnership between Lovelock's hypothesis and elements within the feminist movement is driven by a reaction to this patriarchal imbalance. Most attractive to Gaian feminists is the concept of the Goddess Earth to counterbalance the supposedly male God of the biblical corpus. That God is not conceived of in the first creation story of Genesis 1 as male, but somehow as encompassing both male and female, is forgotten or overlooked. The balance of a male God and a female Goddess is too attractive.

The tie-in with Gaian thought is due to the supposed prominence of Goddess religions, in which the female was equal and often superior to the male, in pre-Judeo-Christian religions. Much is made of excavations at Catal Hüyük, a Neolithic town in the country of Turkey. As a result of excavations at that site, claims are raised that Gaian thought simply recovers age-old truths subjugated and ignored by the patriarchal faiths of Judaism and Christianity, claims that in Goddess religions men and women were equal and all society lived peaceably one with the other.

There is truth to such claims. The fertility religion of Baalism, for example, which the Jews encountered upon entering the Promised Land, did have at least two gods whose copulation supposedly resulted in the ongoing fertility of the land. To claim, however, that males and females lived in equality and that everyone was at peace with everyone else is misleading. Thus, comments such as those by the Gaian author Elisabet Sahtouris when she states that, "it seems that our human childhood which lasted far longer than has our recent adolescence – was guided by religious images of a near and nurturing Mother Goddess before a cruel and distant Father God replaced her in influence"[29] are simply wrong. It might be nice if it were so but the evidence is sketchy at best and inconclusive. Even the noted feminist author Rosemary Radford Ruether attacks such claims noting that "Catal Hüyük society was well supplied with weapons:... There is a clear male-female distinction in associations with material culture; men were buried with weapons,... and women with mirrors, jewelry, and cosmetics." Moreover, she notes, "it is misleading to speak of Catal Hüyük religion as focused solely on a goddess, for the actual shrines show us two major complementary symbols, the Goddess as representative of both birth and death and the bull as symbol of male virility."[30]

More troubling than this lack of historical accuracy concerning the excavations at Catal Hüyük is the fact that if we postulate two gods in the interest of fairness, what happens to the oneness which for me, at least, is the key attraction within Gaian thought? Instead of oneness there is an insolvable dualism. Rather than pursuing the line of thinking which claims that the divine is androgynous, either containing both male and female qualities, or else beyond gender differences, Gaians set up an Earth Goddess to balance the male God of Western religious faiths. This parallels the shift from unisex fashions and the claim that women and men were essentially the same, to the current claim, in many feminist circles, that women and men are different and that women are superior to men. It may be a needed correction, but it cannot be a long-term solution without descending into a matriarchalism every bit as troublesome as the current patriarchalism.

PROBLEMS WITH THE GAIAN THEORY

In spite of a major inconsistency between an emphasis on the interconnectedness and oneness of life and the troubling emphasis on two Gods rather than one, it is clear that the religious implications of Gaian thought form one of its prime attractions. The scientific underpinnings and claims of the Gaia hypothesis may be questionable, but the religious overtones are not. This is not to claim that the religious implications of Gaia theory are benign. I have already mentioned the problem with postulating two Gods and the inevitable dualism that results.

Another key problem area concerns the apocalyptic mentality that grips many Gaians. The displacement of the dominance of the human creature over nature and his/her replacement with worship of the Goddess Ge does not inevitably lead to a gentler, more optimistic religious faith, at least in regard to the human creature. Faced with the polluting ways of human beings, some Gaians hold that the Earth will rise up like some avenging warrior and return the earth to a pre-civilized simplicity. The fact that the majority of the human race would be killed in the process does not seem to be much of a concern. This apocalyptic fervor sounds suspiciously similar to that found within fundamentalist Christianity, specifically within the advocates of premillennial dispensationalism who claim that we are living in the final dispensation and that soon a vengeful God will annihilate most of the human race and destroy the world. It is as ugly and unattractive in Gaian dress as it is in fundamentalist costume!

Moreover, the constant emphasis upon some golden past strikes me as wishful thinking rather than as valid religious hope. The past had its problems as does the present and it is simply naïve to think otherwise. For example, the indigenous peoples of North America had some interesting and helpful religious perspectives which we would do well to recover. Yet to present indigenous groups as benign friends of Mother Earth is likely to overstate the truth. Although Martin Palmer's claim that "in North America 75% of known species of large mammals were hunted to oblivion by the Native Americans before the settlers came,"[31] is wrong; nonetheless, it is true that after the advent of the settlers, the indigenous people of North America participated in the decimation of some species of wildlife.

A helpful religious vision must be able to balance past with future, one of the attractions of the biblical story. I have always been impressed with the fact that while the biblical narrative begins in a rural, garden setting, it does not descend into a simplistic call to turn back the clock. Instead, the thrust is forward to an urban image, the New Jerusalem. It is, though, an urban setting which contains the rural beauty of Eden within it, rather than the urban cities which too many North Americans live in, separated from nature and thus divorced from a wellspring of spiritual nurture and strength. This balance is lacking in Gaian thought; the golden past predominates and the present is depicted in almost totally negative terms.

A final negative feature is the refutation by most Gaians of the Christian doctrine of the resurrection. Most often, this is a misguided refutation, which really attacks the Hellenistic concept of the immortality of the soul rather than the Pauline concept of the resurrection of the body. Even as astute a theologian as Rosemary Radford Ruether falls into this error as she writes, "even as we take into our spirituality and ethical practice the transience of selves, relinquishing the illusion of permanence, and accepting the dissolution of our physical substance into primal energy, to become matter for new organisms, we also come to value again the personal center of each being."[32] The Pauline concept of the resurrection of the body does recognize the transience of the human self. We die, not just our body, but we actually die. However, in the mind of the divine nothing can be lost and so we are reconstituted, resurrected, body and spirit. Of course, Paul was not naïve enough to claim that we come back with the same body; it is a spiritual rather than an earthly body, but it is still a body. Moreover, our resurrection coincides with the liberation of the earth. The problem with Gaian thought is that there can be no personal center of being; all is transient, even Gaia herself; life is the only ongoing reality. The consequences of such a view really negate the insistence upon embodiment that is such an attractive feature of Gaian thought.

STRENGTHS OF THE GAIAN THEORY

Thankfully, the religious strengths of Gaian thought help balance these negative features. These strengths are in helping to integrate the human creature back into the Earth from which, in the biblical creation story, he and she came. The separation of the human creature from nature has resulted in a hubris that has fed the environmental crisis as well as divorced most North Americans from a rich source of spiritual strength. If I had a quarter for every person who told me that they could worship the divine in nature, alone by themselves, rather than in the somewhat artificial constraints of a church building, I could donate my time to my present church for free.

While often such comments are merely an excuse for not working at one's spiritual well-being, many times they are a heartfelt cry for a reunion between the human creature and the created order. We sense instinctively that we are part of the created world, that its future and our future are inexorably intertwined. We know that the health of our souls and the health of the world form a symbiosis which when broken results in the pollution of the environment and the spiritual amputation of the human creature.

This divorce between the world of nature and the human world has fed much of the sexism within our present society. Nature was seen as inferior and since women were connected with nature in a highly visible way – due to the fact of childbirth and menstruation – women were seen to be inferior beings to men. The emotional was subjugated to the rational, imagination to education, poetry to prose. Cold, rational objectivity was treated as superior to emotive subjectivity. Empathy gave way to justice; charity and love gave way to fairness.

The answer, however, to this present crisis in which the gap between the human creature and the creation has grown even larger, is not to pack up your belongings and move to the country, as the writer Sam Keen advocates, but rather to allow Gaian thought to help reintegrate man/woman back into the physical world. We must be converted twice – once from the physical world to the spiritual world and then again from the spiritual world to the physical world. Gaian thinking helps in this necessary process. We need to hear repeatedly the Gaian emphasis upon the circle of life. The interconnectedness of all things provides a much-needed antidote to the specialization and compartmentalization that has afflicted Western society. Gaian theory is important in this task as it takes alternative medicine's protest against the dehumanizing specialization of modern medicine to new levels and broadens it to include the whole earth. In Gaian thinking, no one is an island entire unto themselves, we are all part of life, part of the world. In Christian thought, this could be extended to claim that we are not only part of the world, but also part of the cosmos and, in final instance, part of God. Such a viewpoint would do much to overcome the gender, racial, social, political, and economic differences that plague our world and threaten the stability of our society.

It might also help heal the psychological illness that on a social scale seems to have afflicted Western and North American society. In his book *Hymns to an Unknown God: Awakening the Spirit in Everyday Life*, Sam Keen laments:

I am troubled by the arid character of the modern mind. Something we can only call spirit is drying up, as our underground water sources are being contaminated, exhausted, and dry. I suspect our loss of liquid imagination is related to

*the pollution of our streams and our urban exile. As embodied creatures, what-
ever we do to our environment, we do to ourselves. As without, so within. The
structure of the human soul mirrors the world in which it dwells.*[33]

In a similar vein, Theodore Roszak tells a modern parable of a psychiatrist
with a throng of patients, "exhibiting a veritable compendium of emotional
disorders." One by one these patients are brought in for treatment and du-
tifully obey the psychiatrist's bidding. "They rehearse their dreams, confess
their fears,... but few are cured." Roszak ends with this punchline; "now
step back and view the scene from a wider perspective. The patients report
for therapy to a room. The room is in a building, the building is in a place.
The place has a name. Buchenwald."[34]

If we begin to see ourselves as one with the world, one with others,
then the aridity of the modern mind can be cured and the Buchenwalds
which we have created can be torn down.

I reflected on Roszak's analogy one day and on the concept of the earth
as a self-regulating, living being. I was out for a walk in the beautiful An-
napolis Valley of Nova Scotia. My family had gone away to visit friends and
I was left to continue work on this book. No one was with me except for our
small black poodle, Ruffles. As I climbed the North Mountain near our
house, I looked back over an orchard of apple trees, their blossoms now
gone and in their place small green apples, little more than the size of cher-
ries, beginning to grow. In the distance was the valley with the university
town of Wolfville, the white spire of the Manning Chapel a comforting
point of reference.

Here the valley ends as the Minas Basin, an arm of the Bay of Fundy,
sends its tides to rush in and turn the calm blue water into smooth, brown
milk chocolate until the dirt settles and the blue returns. As I entered a
grove of maple trees punctuated with the odd pine, the view was lost to me.
I stopped, and looked up at the canopy of trees above me which almost but
not quite blocked out the evening sky. The maples felt as if they were talking to each other while the pines slumbered quietly. Was I still in Nova
Scotia in 1997 or was I back in the days when the Mi'kmaq Indians traveled

the forest trails? Perhaps, I let myself imagine, I was in Tolkien's magical Fanghorn forest where the trees take on human form and walk about. This was no Buchenwald but an Eden and oddly enough for someone who has feared being alone, I felt as if I was among friends.

I raised my arms and did a slow dance, a dance of praise to God and unity with the leaves that fluttered and rustled in the breeze. For a moment, I sensed that I was one with the Earth and the trees. I was rooted in the ground as I reached the branches of my arms up high to receive the last of the fading sun and the start of the evening dew. Then the spell was broken. I dropped my arms and wondered to myself, what do we miss when we refuse to see?

That is the story of the next chapter, for the world of the new physics is a world of seeing things in a different perspective. What seems to be self-evident is not self-evident. The world of the senses dissolves into something else, something much more mysterious. I do not understand it all; even a simplified overview is tough going. But, the struggle is well worth it.

3

THE NEW PHYSICS

A university professor was testing a graduating student concerning his knowledge of electricity. "Why and how does electricity work?" the professor asked.

After thinking for a moment, the student responded, "I used to know but I have since forgotten."

"That's too bad," the professor said sarcastically. "Only two people in the whole world know, you and God. And now one of them has forgotten!"

Although probably apocryphal, this joke, which illustrates that humor as well as beauty is in the eye of the beholder, finds its way into many books on science, and particularly physics. When specialists themselves joke about the unintelligibility of their material, then the layperson had better be wary. If you are tilted back in your easy chair with your slippers on and a cup of coffee in your hand, wanting to read a bit before you turn in for the night, you might want to skip this chapter and move on to chapter four. However, I hope that you will come back to it some time because just as theology is too important to be left up to the theologians, so science is too important to be left up to the scientists. In spite of the difficulty of the material, it affects our lives and influences our views so powerfully that it is foolhardy and dangerous to blind ourselves to the world of science. Moreover, alternative medical supporters, Gaian advocates, New Agers, near-death experiencers, even the new breed of Pentecostals, point to the insights of the new physics to support their perspectives and practices.

Consider, for example, the formidable effect Newton's views have had and still have upon our thinking. Without even realizing, we begin, as a result of Newtonian science, to view the world as an elaborate machine that can be broken down into its constituent parts and thereby be manipulated and controlled. We think of space as a vast dark emptiness punctuated here and there by stars, some of which have planets like our own that circle around the star, held in place by invisible strings of gravity.

And God? Well, if God has a place at all in this world view it is as the "Unmoved Mover," the one who sets everything in motion and then sits back and watches. Little wonder that in spite of Newton's own intense religiosity, the effect of his scientific understanding further secularized Western wo/man as prayer and miracles came under attack as naïve and prescientific holdovers from the past. What use was prayer if the world operated according to fixed and immutable laws? How could miracles take place without being disruptive of the natural order? Did a personal, loving God who related intimately to men and women make any sense in the Newtonian world view? Increasingly, many people concluded that the concept of a personal God made little sense in light of modern science. St. Anselm's method of "faith seeking understanding" was abandoned as leading to a dead end, while the small gap between Thomas Aquinas' world of nature and supernature became an unbridgeable chasm. Modern men and women were left with the option of becoming fideists (I believe, even though it is absurd) or atheists (there is no God). There seemed to be no other reasonable option.

In the late 1960s, however, a change began to take place. Suddenly, science and religion were brought back into dialogue and sometimes into partnership. This was signaled by the publication in 1975 of the book *The Tao of Physics*. Written by a relatively unknown physicist, Fritjof Capra, this book trumpeted the similarities between the new discoveries in physics and ancient views of the world found primarily among adherents of Hinduism and Buddhism. *The Tao of Physics* caused little commotion among scientists and received its warmest reception among theologians, particularly those connected with the new academic field of religious studies.

Capra's interests bore fruit in the 1980s as leading physicists began to write about themes formerly considered to be within the domain of religion rather than science. Thus, Paul Davies, an Australian scientist and recent winner of the $1 million 1995 Templeton Prize for progress in religion, has published a variety of books on the new physics with such provocative titles as *God and the New Physics* and *The Mind of God*. And, suddenly, everyone's coffee table seemed to sprout Stephen Hawking's book *A Brief History of Time*. In this bestseller, Hawking, who holds the chair of Lucasian Professor of Mathematics at Cambridge University (formerly held by Newton himself), muses about the possibility of discovering the Theory of Everything (TOF) or the Grand Unified Theory (GUT). He concludes:

...if we do discover a complete theory, it should in time be understandable in broad principle by everyone, not just a few scientists. Then we shall all, philosophers, scientists, and just ordinary people, be able to take part in the discussion of the question of why it is that we and the universe exist. If we find the answer to that, it would be the ultimate triumph of human reason for then we would know the mind of God.[1]

The scientific theory that has sparked this interest is not new, at least not to the scientific community. Quantum theory was established by Max Karl Ernest Planck in 1900, the theory of relativity by Albert Einstein in 1905, and quantum mechanics in 1926 by Max Born and Werner Karl Planck. However, the implications of what is commonly called the "new physics" have taken a long time to penetrate the world view of the layperson. In part, this has been due to the specialization of knowledge, in part due to the difficulty of picturing the quantum world or conceptualizing the theories of relativity. In an editorial in the *New York Times* written on January 28, 1928, the author complained,

The new physics comes perilously close to proving what most of us cannot believe; until we have rid ourselves completely of established notions and forms of thought. Relativity translates time into terms of space and space into terms of time. The quantum invites us to think of something which can be in two places at the same time, or which can move from one spot to another without passing through intervening space.

[The editorialist concluded] In this turmoil, there is at least one possible source of comfort. Earnest people who have considered it their duty to keep abreast of science by readapting their lives to the new physics may now safely wait until the results of the new discoveries have been fully tested out by time, harmonized and sifted down to a formula that will hold for a fair term of years... Reshaping life in accordance with the new physics is no use at all. Much better to wait for the new physics to reshape our lives for us as the Newtonian physics did.[2]

It is probably no coincidence, then, that when quantum mechanics began to reshape our world through the computer chip – since the quantum world deals with the very small, the level at which a computer chip operates – interest in the wider implications of the new physics began to take root, in spite of the fact that it is no easier to visualize in the 1990s than it was in the 1920s.

This inability to visualize many of the axioms of modern physical theory is due mainly to the abstract nature of modern physics, which has outstripped the ability of our senses to verify, even when magnified by machines. Einstein was once told that a researcher had verified one of his theories through an experiment. When asked how he would have felt if the experiment had disproved the theory, he supposedly replied, "so much worse for the experiment!"

While the role of theory and experimental verification is like the old conundrum of the chicken and the egg, in the past the two were always closely allied. Observation led to theory, which led to experimental verification. The classic and probably fictitious example of this is Newton sitting under his apple tree. The apple drops to the ground, hitting Newton on the head, and sparking the theory of gravitational attraction. This theory is then tested and confirmed through experimental verification, or the experiment shows the theory to be false and in need of modification. In which case a new theory is then proposed, tested and accepted as truth or not.

One way of defining how the "new physics" differs from the "old physics," then, is through the dominance of theory (expressed in mathematical formula) over experimental verification. This leads to two distinct groups of scientists fed by two very different philosophical understandings. The first group feels that nothing much has changed except that physics now deals with the world of the very small (quantum world) or the very large (cosmology) and that in these realms, because of the crudity of our senses even when extended artificially, experimental verification is difficult and may take centuries to access. However, the method of observation leading to theory which is then confirmed by experiment still holds true. Thus, Roger Penrose insists that the new and bizarre world of quantum theory simply results from the prior process of observation.

How do we know that classical physics is not actually true of our world? The main reasons are experimental. Quantum theory was not wished upon us by theorists. It was (for the most part) with great reluctance that they found themselves driven to this strange and, in many ways, philosophically unsatisfying view of a world.[3]

Stephen Hawking echoes this perspective in his book *A Brief History of Time,* claiming that a good theory is based on observation and "must make definite predictions about the results of future observations."[4]

However, as the astronomer David Lindley notes, "experimental physics are at the point of impossibility,"[5] "to go further we must deliberately close our eyes and trust to intellect alone."[6] This necessity of trusting to intellect alone has given rise to the second group who, rather than asking for our patience while observation catches up to theory, take a different tack and conflate theory and observation.

This was the approach taken by the founders of quantum theory. In opposition to Einstein who held that there was a "real" world out there that was governed by natural laws, physicists such as Niels Bohr preferred to remain skeptical about some objective reality. Instead, reality emerged out of and in response to the act of measuring. This was brought home dramatically by Werner Heisenberg's uncertainty principle, which states that one cannot measure two complementary aspects of something. If, for example, you wish to measure the location of a rolling golf ball you do so at the expense of its motion. If you wish to measure its motion, you do so at the expense of its location. In the world of classical physics, it seems self-evident that this can be done. In the world of quantum physics, it is impossible; the act of measuring affects that which is being measured.

The difference between the two groups leads many writers to categorize the new physics as consisting only of quantum mechanics. Einstein's two theories of general and special relativity belong to the world of classical physics, such people claim, since Einstein held to an objective reality which, given time and technological advancement, could be measured objectively. Thus, Gary Zukav, in his book *The Dancing Wu Li Masters: an Overview of the New Physics*, contradicts himself. First, he claims that the new physics is the physics of quantum mechanics "which began with Max Planck's theory of quanta in 1900, and relativity, which began with Albert Einstein's special theory of relativity in 1905."[7] However, he then immediately proceeds to claim that classical physics includes the theory of relativity but not quantum theory, since the theory of relativity was based on a belief in an objective reality, which could be experimentally verified!

When the emphasis is placed on the dominance of theory over experiment, however, it becomes clear that the "new physics" must include both Einstein's theories of relativity as well as quantum mechanics. In both the quantum world and in Einstein's theories of relativity, academic theory and experimental verification are no longer in close partnership; theory has separated from sensory confirmation.

This divorce between theory and observation is not true of traditional physics, commonly known as Newtonian or classical physics. Sometimes theory preceded observation, sometimes observation preceded theory, but always the two were tied together like a horse and carriage, or like love and marriage as the popular song puts it. In Newtonian physics cause and effect go hand in hand, theory is confirmed by sensory observation and the world is conceived of as a gigantic machine which can be broken down into its constituent parts and put back together again.

CHALLENGES TO THE
WORLD VIEW OF CLASSICAL PHYSICS

The breakdown of Newton's classical synthesis began as a result of the work of a British physicist, James Clerk Maxwell, born in Edinburgh, Scotland, in 1831. In a series of papers published in the 1860s, he analyzed mathematically the theory of electromagnetic fields and predicted that visible light was an electromagnetic phenomenon that traveled at an invariant speed. This observation had two important effects. The first dealt a deathblow to Newton's theory of light as made up of particles. Light was now seen to travel through space in a wavelike motion in company with other kinds of electromagnetic waves, which today include radio and television waves, radar, microwaves, and X rays. By extension, it was also influential in Albert Einstein's construction of his theory of special relativity in 1905. According to Maxwell's theory, the speed of light remained the same for someone at rest or in motion. This seemed ridiculous to most, if not all, of Maxwell's contemporaries. Relative speed was a well-established principle as any child could tell you who has had to chase down a runaway cap blown this way and that by a winter wind. Even when Maxwell's prediction was proved correct in the 1880s as the result of experiments done by American physicists Albert Michelson and Edward Morley, scientists did not know what to make of it. The young German physicist Albert Einstein did, however. He took Maxwell's prediction at face value; the speed of light was invariable. It led him to construct a fourth dimension known popularly as space-time.[8]

This fourth dimension of space-time is inconceivable apart from mathematical theory. We can imagine what space is, and we think we know what time is, but, try as we might, space-time is unimaginable except by using mathematical theory. The same holds true for the general theory of relativity, which postulates a new way of understanding gravitational attraction based on the curvature of space rather than on invisible strings of gravity. The old clay and Styrofoam models of our galaxy that almost all of us made in high school have been shown to be inadequate. At large scales, gravity acts differently than it should according to Newton's calculations. Rather than the clay and Styrofoam balls, a better representation of the universe would be a rubber sheet indented by the sun. The planets are caught not by the gravitational pull of the sun, but by the curvature of space. Like a marble on a roulette wheel our Earth spins around the sun, but without friction it never slows and thus never falls into the center. Even this is a poor representation of the theory of general relativity, which like special relativity is unimaginable except through mathematics. Thus, the modern physicist has learned to distrust common sense and intuition, and to look instead to mathematical and logical consistency for a guide into the dark and distant places where the frontiers of physics now press.

This trust in the power of the intellect and the primacy of theory is particularly noticeable in the world of quantum mechanics and especially quantum cosmology. As a result of his work with black body radiation, Max Planck found that the classical laws of Newtonian physics were inadequate to describe the effects he was observing. According to classical theory, as temperature increases the atoms comprising that which is being heated vibrate more rapidly and thus radiate more energy. The correlation between temperature and radiation should therefore be a straightforward one. It proved to be nothing of the sort. Rather than study the phenomenon more closely to see what was going on, Max Planck, almost magically, came up with a mathematical formula that exactly described what he was observing. Then, moving from theory to physical reality, Planck deduced that energy can only be radiated in quanta of energy hv where v is the frequency and h is the quantum action, now known as Planck's constant. Perhaps the best illustration of this is to compare the musical frequencies of a violin with those of a piano. A violin can produce a continuous spectrum of pitches or frequencies, a piano cannot. A "B" on a piano is 494 oscillations per second and a "B-flat" is 466. Since the piano is quantized in a way that a violin is not, the pianist cannot play a note of 488 oscillations per second, while a violinist can.

The important point in all this is that while sound waves are vibrations in the air which can be sensed, or heard, (within certain limits, of course) and light waves are vibrations on an electromagnetic field which can also be measured, or seen, (again within limits) quantum waves are "a purely mathematical abstraction."[9] They cannot be measured; they only give probabilistic data.

The union of quantum theory (which applies only at distances of atomic dimensions) with cosmology, helps to explain what for me was a very puzzling problem. One of the key physicists working in the area of quantum cosmology is the physicist Stephen Hawking, who suffers from a crippling illness known as Lou Gehrig's disease which confines him to a wheelchair, scarcely able to move. "How," I asked myself, "could someone unable to conduct even the most rudimentary of physical experiments be able to be on the forefront of physical theory?" The answer, of course, is that physical theory is just that – theory. It is no longer based on observation but on mathematical constructs. The intellect has triumphed over the senses. Ideas dominate over experiments.

All this serves to explain why the two fields of religion and science, long thought to be concerned with separate areas of knowledge, have now begun to bump up against each other. Religion has always dealt with subjects that were beyond the physical senses to prove or disprove. One cannot prove the existence of God, for example, except tangentially. Direct proof of God is an impossibility. As physics enters the realm where direct proof is impossible and only tangential, indirect proofs are available; the partnership between religion and science, where religion concerns itself with the "why" and science with the "how," no longer holds true. Thus, in his book *God and the New Physics*, Davies notes that the theories of relativity and quantum theory have forced physicists to approach physics in "novel ways that seemed to turn common sense on its head and find closer accord with mysticism than materialism."[10] On a more popular level, the author Nancy Logan makes this interesting confession:

I still remember my grade 11 physics class. I learned two world-shattering things: One. Chuck Joiner, a gorgeous blond I'd loved from afar for two years, turned out to be as dumb as a bag of hammers. Two. The entire universe is made up of matter and nothing but matter. Both discoveries shook my 15-year-old life. But the first I'd gotten over by grade 12 when I developed a crush on my chemistry teacher, Mr. Laver. The second shaped the next two decades of my life.

If nothing but matter matters, what place is there for the mind? Noble pursuits like poetry and art? And what happens when we die?... If matter is eternal, just chugging blindly along on its own, what place is there for God and the human spirit?

These depressing questions were never answered. Not by my university courses. Not by my church... Science had stolen my God...

And then, a few years ago, I stumbled upon some tantalizing tidbits from "the new science," suggesting something behind the physical, behind the chunks of matter whirling through the universe. I found talk of spirit and mind and beauty. And talk of God. [emphasis mine][11]

This same confluence, particularly as it has manifested itself in the field of quantum cosmology, prompted Robert Jastrow in his short survey of the interaction between the new physics and cosmology to claim that "for the scientist who has lived by his faith in the power of reason, the story ends like a bad dream. He has scaled the mountains of ignorance; he is about to conquer the highest peak; as he pulls himself over the final rock, he is greeted by a band of theologians who have been sitting there for centuries."[12]

SCIENCE AND FAITH

The early encounters between science – as developed in Greece – and Christian faith were marked by respect. While the North African theologian, Tertullian, held that there could be no relationship between Athens (the center of culture) and Jerusalem (the center of faith), he was in the minority. The majority respected the intellectual achievements of Greek society and even spoke of Greek philosophy as an anticipation of Christian theology, as pre-evangelism, and depicted Socrates as a prefiguration of Jesus Christ.

This affection for Greek philosophy among Christian theologians gave rise to the almost total domination of Platonic thought during the Middle Ages. Using his famous analogy of the cave, Plato taught that the real world to which the true thinker must aspire was the world of ideas rather than the world of the senses. It was this fascination with Platonism (or rather neo-

Platonism as mediated by Plotinus) rather than any Christian antagonism toward *science,* which accounts for the slowness of scientific development during this time. Indeed, when science *did* begin to establish itself as a separate field of endeavor it was because the *theologian* Thomas Aquinas introduced the thoughts of another Greek, Aristotle, to the Western world.

Although Aristotle was a student of Plato and carried on much of Plato's work, his most important legacy was to reverse Plato's interest in timeless ideals which could not be approached by the senses but by the intellect alone. For Aristotle, knowledge obtained through the senses was *important* and could lead in a logical procession to these ideals. Through the writing of Thomas Aquinas, this led to the division of the world into two sections. The world of nature would be studied using sensory data. In the world of the supernatural, faith would supplant the senses. In this way, the Christian church both preserved its Platonic heritage and welcomed the new understandings ushered in by the rediscovery of Aristotle.

THE SEPARATION OF SCIENCE AND RELIGION

This partnership, forged in the 13th century, lasted until the 18th century and the rise of the movement known as the Enlightenment. The scientific breakthroughs of Isaac Newton in the 17th century (and later the scientific method of Francis Bacon, wherein the meaning of the whole was to be found by breaking it down into its constituent parts) helped pave the way for the Enlightenment's attempt to change the relationship between science and religion. It wasn't until the German philosopher Immanuel Kant, however, that the relationship between science and religion began to break down. At most, Newton's and Bacon's work made the relationship between religion and science one of equals. No longer was religion the culmination of science. Instead, the two existed side by side. The book of nature, as it was often referred to, and the book of the Bible were two complementary ways to understand God. But, while Aquinas felt that religion was not irrational or antirational (even if it was at times supra-rational), Immanuel Kant taught that religion had nothing to do with reason. Intellectual argument could not lead one from the world of the natural to the world of the supernatural in a continuous fashion. According to Kant, the finite could not comprehend the infinite. The starry sky above and the moral law within the sense of wonder and the sense of moral "oughtness" led to God, but not the mind – the mind was trapped in the prison of the world, much like the senses.

Kant's writings, then, had the result in some circles of severing religion from science. No longer equals, they became competitors, antagonists, enemies. Religion began to recede while science and its partner technology grew in influence and in popularity. Today, while religion maintains a place in the affections of people, it is a humbled and reduced sphere of influence. The priests of our age are the scientists. As Roger Jones, a physicist at the University of Minnesota puts it, "even in the popular imagination, science tends to dominate. It isn't only that science and technology are pervasive on television and in film and in literature. More significantly, our very conceptions of the world are strongly colored by science." He concludes, "science is no longer merely a field of study – an 'academic discipline.' In our culture, it has become a way of life and a system of belief."[13]

The high-water mark of science's separation from religion can be seen in the work of the French scientist, Marquis de Laplace. At the beginning of the 19th century, Laplace argued that the universe was completely deterministic and that there was a set of scientific laws which would allow the scientist to predict and presumably manipulate the universe. According to such a view, God is no longer necessary. Scientific laws have taken God's place, the redeemer is not the Christ of the Bible but the human creature, and the Holy Spirit is displaced by a belief in inevitable progress.

It is when measured against the Laplacian world of determinism and inviolable natural laws which can explain everything, that the new physics of relativity and quantum theory appear so religious. The Laplacian hubris, however, is still alive in the minds and works of people such as Frank Tipler, the late Carl Sagan, Stephen Hawking, and others. For example, Francis Crick, in his book *The Astonishing Hypothesis: the Scientific Search for the Soul*, concludes that "other hypotheses about man's nature, especially those based on religious beliefs, are based on evidence that is even more flimsy but this is not itself a decisive argument against them. Only scientific certainty (with all its limitations) can in the long run rid us of the superstitions of our ancestors."[14] The mathematician, Frank Tipler notes that not only must religion bow down before the claims of science, but also even scientific disciplines such as biology will fall before the domination of physics!

Galileo got into serious trouble when he dared to invade the turf of the philosophers and theologians. However, as I implied by my earlier description of the domain of physics – the whole of reality – an invasion of other disciplines by physicists is inevitable, and indeed the advance of science can be measured by the extent of the conquest of other disciplines by physics.[15]

The majority of scientists and interested observers, however, realized that something was different with the new physics. In Einstein's case it may simply have been his penchant for using the word "God" when really he meant some sort of orderly universe that had little to do with the God of theism. However, it shouldn't be overlooked that it was Einstein who, reversing his original position that the universe was stable and unchanging, popularized the concept of an expanding universe. An expanding universe, by logical deduction, necessitates a beginning in time and space which, in turn, leads to a creator who is intriguingly close to the creator God of the biblical drama.

With many of the founders of the quantum theory, this religious bent was even more pronounced. Max Planck, for instance, held to a fairly orthodox Christian faith while Werner Heisenberg has written extensively on the role of religion and science, relating a key moment of mystical illumination which took place when he was a young man attending a youth assembly in the Altmühl Valley in Germany. Confused by the way atoms bond together and intrigued by Plato's writings on the subject, Heisenberg writes of hearing a violinist play the first few notes of Bach's *Chaconne*:

All at once, and with utter certainty, I had found my link with the center. The moonlit Atmuhl Valley below would have been reason enough for a romantic transfiguration; but that was not it. The clear phrases of the

Chaconne *touched me like a cool wind breaking through the mist and revealing the towering structures of the beyond. There has always been a path to the central order in the language of music, in philosophy, and in religion, today no less than in Plato's day and in Bach's. This I now knew from my own experience.*[16]

RESPONSES TO THE NEW PHYSICS

One response to the religious and philosophical implications of quantum theory has been to stress the mystic overtones of many of the discoveries in this new field of science and to tie these in with Eastern philosophies and religions. Fritjof Capra was the first in what has been a long line. Many, if not all of this group of writers and thinkers, have been connected with the Esalen Institute founded in 1962 at Big Sur, California. One such participant is Richard Tarnas who at one time was Director of Programs and Education at the Esalen Institute. Writing about the reaction that he and other like-minded people had to the insights of the new physics he says, "to many who had regarded the scientific universe of mechanistic and materialistic determinism as antithetical to human values, the quantum-relativistic revolution represented an unexpected and welcome broaching of new intellectual possibilities. Matter's former hard substantiality had given way to a reality perhaps more conducive to spiritual interpretation."[17]

This spiritual interpretation was always couched in Eastern religious terms (particularly the tenets of the Buddhist faith). It was also part of a larger mix of themes such as the recovery of the feminine perspective in history, religion and science; the importance of the environment; and most recently, enthusiastic acclaim for the Gaia theory of a living earth. As well, there has been a growing fascination with holism as an antidote to the increasingly fractured society of post-World War II North America. A new vision was supposedly emerging which stood in sharp contrast to former ways of thinking in both science and religion. This new vision has been articulated in various ways and one can find various charts sketched out in the writings of Fritjof Capra[18] as well as in Marilyn Ferguson's seminal book on the New Age movement, *The Aquarian Conspiracy*.[19]

My own analysis of this new vision is that it stresses the fluid over the static, the universalistic over the dualistic, the circular over the linear, holism and interdependency rather than specialization and independence. It delights in the uncertainty principle of Werner Heisenberg; in the "spooky action at a distance" as found in Bell's theorem, where two particles although separated from each other continue to act in tandem, a fact which troubled Albert Einstein. It resonates with chaos theory, which holds that a butterfly flapping its wings in China affects the weather in downtown New York. It is particularly fascinated with David Bohm's work with holograms. Unlike other visual images where, if you tear off a piece, the torn piece is only an incomplete part of the once large whole, holograms can be ripped into pieces and yet when a laser beam is projected through the torn fragment what is seen is not just a piece of the whole but a complete miniature version of the whole.

Those who have used the insights of the new physics in this manner have not been without their critics. David Lindley, writing about the popular response to the new physics, particularly as mediated through participants in the Esalen Institute, notes, "modern physics,... is popularly understood to have done away with the essential classical ideal of an objective world existing securely out there, waiting to be measured and analyzed by us... Determinism is blown away, free will restored, history given back its arbitrariness and humanity its spontaneity." The reality, Lindley states, is very different from this popular portrayal. Quantum mechanics and relativity have changed some basic notions on which the classical physics of Isaac Newton was built. Thus, "it is true that under the new laws one has to take account of the measurer in order to define and understand the thing measured, but there are rules for doing this accounting; it is not whimsical."[20] Lindley's interpretation of the new physics is echoed by Ian Barbour, a well-known academic on the interaction between science and religion who writes, "probability waves may seem less substantial than billiard ball atoms, and matter that converts to radiant energy may appear immaterial. But the new atom is no more spiritual than the old, and it is still detected through physical interactions."[21]

Most of these authors call for a continuation of the old arrangement whereby science had its field of expertise and religion had another. Some authors want to avoid the situation whereby God is fitted into the gaps in scientific thought, gaps which have had an embarrassing tendency to be filled in by science. For others, less committed to a religious faith, the motive is probably not to preserve a faith in a divine being, but to continue to allow science the domination it has experienced in the Western world for the last two to three hundred years. Thus, in regard to the former motive, Stanley Jaki, a Roman Catholic priest and a scientist, warns that while "there might be many gaps in the present-day scientific information about the electron,... none of these gaps can serve as a doorway to God."[22] In fact, Jaki seeks to rule out any use of science to confirm theological truths, asserting that, "above all, the theologian should be aware of the danger inherent in any quasi-scientific attempt at casting physics in the role of a theological timekeeper."[23] While in regard to the second motive, Frank Tipler probably has it right when he states,

Most contemporary scientists agree... that science and religion can have nothing to do with one another. For example, the council of the U.S. National Academy of Sciences decreed in a Resolution dated 25 August 1981: "Religion and science are separate and mutually exclusive realms of human thought whose presentation in the same context leads to misunderstanding of both scientific theory and religious belief." What scientists who make statements like this really mean is that religion is emotional nonsense, expressing nothing but our fear of death and the primitive view that the natural world is animate. Such scientists regard any attempt to fully integrate science and religion as a reactionary throwback to a prescientific model of reality.[24]

There are, however, those who feel that science will validate religious claims, and that true religion has nothing to fear from science. The problem is not science, but the fact that incomplete and provisional theories are propagated and received as complete and fully formed beliefs. Thus, Ralph Wendell Burhoe, one of the chief architects of the Center for Advanced Study in

Religion and Science, which for the past 30 years has produced the academic journal *Zygon*, admitted in 1987 that he had been motivated by a dream, a dream in which "if one looked at religion in the full light of today's much more advanced sciences, rather than as merely a phenomenon not examinable by them,... one would find that the basics of traditional values not only were *scientifically* valid but, exactly because of this, were more than ever *religiously* true and compelling." On that occasion of the 20th anniversary of the journal's existence, Burhoe felt that finally victory could be claimed in the assertion that science proved religion.[25]

CONCLUSIONS

There are, therefore, at least four main responses to the impact which the new physics has in regard to religion. The first is that science will finally swallow up religion, that a Grand Unified Theory (GUT) or a Theory of Everything (TOF) will be found by which we can explain everything scientifically and God will become redundant. To use Stephen Hawking's phrase, once we discover the Grand Unified Theory we will then, "know the mind of God." The second is that the new physics supports religion, but only in an Eastern dress and that the new physics is part of an exciting paradigm shift that has gripped Western society and beyond. The third is that physics and religion deal with separate fields, using separate methods, which should not be confused or combined. The fourth is that the new physics proves the validity of religious claims, particularly in regard to values and ethics.

The truth is somewhere in between these four assertions. It is clear that the new insights of contemporary physics have dramatically affected the arrogance of science, inserting a much-needed dose of humility. Whether one wants to say that the feminine emphasis has tempered the masculine, Promethean mentality, as Richard Tarnas does in his bestselling book *The Passion of the Western Mind*, or simply assert that the Newtonian scheme has been shown to be inadequate in regard to the very small and the very large (the world of quantum theory on one hand and relativity theory on the other), scientists have now been made aware of mystery, uncertainty, and the impossibility of separating the subjectivity of the observer from scientific experimentation. In fact, one religious writer questions whether the reunification of science and religion can begin by finding a common ground in "not-knowing."[26]

It is also clear that some of the new scientific insights have fascinating analogies in religion. For example, the Heisenberg principle, which states that two complementary properties cannot be studied at one and the same time and yet both are true, finds an echo in traditional Christian assertions concerning the deity and the humanity of Jesus of Nazareth. The earliest creeds asserted the full deity and the full humanity of Jesus. From a Western, rational viewpoint both cannot be true and so various attempts have been made to disprove one or the other of these statements, with the predominant tendency in North America being an emphasis on the humanity of Jesus. Religious thinkers from many different traditions have long known that two paradoxical statements can both be true, but it seems that science has just begun to find this out.

Another fascinating analogy came out of the famous EPR experiment of Albert Einstein, Boris Podolsky, and Nathan Rosen. The idea behind the EPR paper was to come up with a thought experiment which could be done in principle, but whose outcome could not be predicted by quantum theory. By doing so, Einstein hoped to prove that quantum theory was incomplete and that it was merely a stepping stone to another, more satisfying view of the world. The exact opposite proved to be the case. Einstein attempted to measure indirectly two complementary aspects of an electron and thus disprove one of the cornerstones of quantum theory – Heisenberg's uncertainty principle. He did this by using the theory of the conservation of momentum, which was common to classical and quantum theory.

For some time there was no way of knowing whether or not Einstein had succeeded. Then the British physicist Albert Bell conducted an experiment which showed that if you measured two particles which had once been in contact with each other but had been split apart, in measuring one particle you ended up changing the other particle. Since nothing moved faster than the speed of light, how was it possible that particle "a" could affect particle "b"? The answer was to give up the notions of causality and objectivity which Einstein treasured and to opt for the interpretation put forward by Niels Bohr. Bohr claimed that the inner reality of something could never be established, only the predictability of its behavior.

Bell's findings encouraged religious thinkers who felt that his experiment opened the way for prayer and spiritual healing to be both religiously true and scientifically accurate. Somewhere there was some form of energy which moved faster than light. This energy, for lack of a better word, was the arena in which the divine worked, the way in which prayers were answered and miraculous healings took place.

A third analogy is found in the unprovability of many of the newest scientific insights. This has already been mentioned, but it bears repeating. The divine cannot be scientifically proved through traditional sensory means. But then neither can many of the accepted insights of relativity and quantum theory. We know that they work, but then we also (at least some of us) know that the divine works as well. We feel God's presence; we sense God's touch. Since simplicity and beauty have become essential ways in which, in the absence of sensory validation, scientists pick and choose between good theories and bad theories, it is fair to ask whether the same criteria can also be applied to faith.

A fourth analogy arises out of the fact that the major advances in the new physics all have had to do with some aspect of light. First, there were the energy-carrying fields of Maxwell's electromagnetic theory that gave rise to the crucial role light plays in Einstein's special relativity. Second, the tiny differences between Newton's theory of gravity and Einstein's general theory of relativity only become significant at speeds approaching the speed of light. Third, the wave-particle duality of quantum theory was first observed in the pattern of light, which sometimes acted as a particle and sometimes as a wave. What I find intriguing about this is that the Bible is very reticent to describe God. Even in the Hebrew Scriptures (Old Testament), where anthropomorphic characteristics are commonly assigned to God, there is great caution. Moses is not allowed to see God's face on Mount Sinai. The sacred name of God, "Yahweh" is not to be spoken aloud. The only physical descriptions of God are found in the final book of the Bible where

God is depicted by various colors and lights (Revelation 4:2-3 NRSV). This brings to mind Einstein's famous formula that energy equals mass times the speed of light squared. Moreover, once we get to the chapter on near-death experiences we will find those who claim that the dead cannot be seen *because they move at speeds faster than light.* In other words, light becomes the boundary – and perhaps the point of intersection – between the physical world and the spiritual world.

These analogies give hope for a new partnership between science and religion which can overcome both the antiscientific stance of the fundamentalists *and* the capitulation to a materialistic scientific world view which characterizes most mainline Christians. Whereas before, a commitment to scientific thought forced a radical discontinuity between the world of the spirit and the world of the material, no longer is this the case. The evidence is growing that the spiritual operates in harmony and not in antagonism with the material.

All this may be labeled wishful thinking by the scientist still entrenched in a classical, Newtonian position. Nevertheless, at the very least, it is fair to claim that the future will see an exciting and energetic dialogue between the world of science and that of religion. The dividing walls of the past are thrown to the ground and the way is open for the first time in centuries to an exciting new future!

PART TWO

THE SPIRITUAL TERRAIN

4

THE NEW AGE
MOVEMENT

My first exposure to the New Age movement took place at a gathering of clergy in the university city of Kingston, Ontario. A United Church minister whom I knew casually ended up sitting next to me for lunch. We got talking about common interests, the church he had just resigned from, the local theological college which was attached to Queen's University, when suddenly the conversation changed and he began to grow intense.

"You know you have God inside you," he said to me in a low voice.

"I know," I replied, wondering what he was getting at.

"You have Christ inside you," he continued.

"Yes," I responded, more in the form of a question than a statement. And then he delivered his concluding statement:

"In fact, you are Christ. I am the Christ. Each of us is the Messiah."

I am a bit thickheaded but I finally began to realize that this United Church minister was trying to convert me to the New Age movement. "Heh, heh," I interjected. "Wait one minute. You're talking to a good old-fashioned Baptist and I know who the Messiah is and it sure ain't me." At that my ministerial colleague backed off and we continued on a more superficial line of conversation.

I sometimes regret that I put this person off and refused to discuss New Age thought with him. After all, on one level what really is the difference between the remarks of people such as my minister friend and the comments of the early church writer St. Athanasius who asserted that, "the Word was made man in order that we might be made divine?"[1]

There has always been a strong mystical tradition in both Eastern and Western Christianity that would be comfortable speaking in New-Age-like terms about the deity of the human creature. In fact, the presence within the Christian tradition of the best of New Age thinking makes me wonder *why* this minister would want to abandon Christianity for the New Age movement. To make my position clear at the start, most of what I find attractive within the New Age movement is already part of the Christian heritage.

ANCIENT ROOTS OF THE NEW AGE

The New Age movement should be dated from the 1960s at the earliest, and more likely the 1970s; but important, historical foreshadows of New Age perspectives stretch back to the Gnostic heresy of the second and third centuries. Gnosticism was one of the early options which Christianity could have taken but chose not to. Paralleling the mystery religions, which were extremely popular at the time, Gnostics interpreted Christianity as another, superior mystery religion. Salvation was not achieved through faith in the work of Jesus, but rather through the possession of secret knowledge known only to the initiates.

The chief elements of this secret knowledge were twofold. The first was that the god of the Old Testament (the god of creation) and the God of the New Testament were different beings. The god of the Old Testament was a lesser god, known as the Demi-urge, who created an inferior, physical world. The God of the New Testament was the primary God, the God of the Spirit, who created man and woman to be divine, spiritual beings. The task of the believer was to liberate the divine being trapped within each person and to do so through ever increasing one's knowledge, both of God and of the destiny of humanity, as well as through participation in esoteric rites.

I have always found a troubling contradiction between the dualism of Gnostic teaching which treats mind and spirit as superior to flesh and matter, and the Earth-affirming position of many New Agers, particularly those who believe in Lovelock's Gaian hypothesis and who feel some affinity with neo-paganism as manifested in the Wicca religion. New Agers themselves, though, fail to draw such conclusions. Instead, within the New Age world, Gnosticism has, as Martin Palmer puts it, "become something of a fad." Palmer suspects that this is because Gnosticism "is seen as offering a form of Christianity which was suppressed and therefore must be valid; and it proposes a world view in which those who are in the know are superior to those without the knowledge."[2] Such a view seems a bit too churlish to me. I would suggest that the New Age movement, like other new religious movements, has sought to legitimize itself by creating a history that can compete with traditional Christianity.

While Gnosticism itself died out in the fourth century, Gnostic dualistic ideas were picked up by the Manichean movement which persisted until at least the 13th century, as well as by the Cathari and Albigensians, two medieval European sects. However, all three of these groups focused on the dualistic teachings of Gnosticism while the New Age movement emphasizes only certain aspects of Gnostic thought, specifically the divinization of the human being and, to a lesser degree, the emphasis upon salvation through the possession of secret knowledge.

Gnosticism, however, plays only a small part in New Age thought. Another Christian teaching known as millenarianism plays a much more significant role. Millenarian thought is a direct outgrowth of the apocalyptic literature of both the Hebrew Scriptures, specifically the Book of Daniel, and the New Testament Book of Revelation. Apocalyptic thought adopted a dualism of the present and the future, as opposed to the flesh/spirit dualism of Gnosticism.

Millenarianism took its name from the belief that there would be a thousand-year reign of the saints on Earth either before or after the return of Christ as predicted in Revelation 20:1–7. Because of the increasing tendency of millenarianists to rhapsodize over the physical pleasures of the saints during this thousand-year period, church leaders such as St. Augustine turned against millenarianism. In the 12th century, though, millenarian teachings resurfaced under the leadership of a Cistercian Abbot named

Joachim De Friore who added a new twist in that he postulated three ages. The first age was the Age of the Father and corresponds to the historical period covered by the Hebrew Scriptures. The second, the Age of the Son, according to De Friore, began with the birth of Jesus and was to end somewhere around the beginning of the 13th century. At that time, a third age, the Age of the Spirit, would arise.

De Friore never pushed his views to the point of displeasing the Pope, who had a vested interest in the Age of the Son continuing and an antipathy to the radical spirituality that would arise during the Age of the Spirit. De Friore's vision, however, proved to be influential in the rise of the radical wing of the Anabaptist movement under the leadership of Thomas Müntzer. Müntzer sought to inaugurate the new age of the Spirit by allying himself with the Peasant's Revolt in Germany in 1524. This was too much even for the Protestant followers of Martin Luther, who cooperated with the Roman Catholic church to get rid of this troublemaker.

After Thomas Müntzer, millenarian thought took on a secular, revolutionary character. This more secularized form of millenarianism influenced people as diverse as Oliver Cromwell and Karl Marx.

Again, we find a strange contradiction. The New Age movement has picked up the millenarianist vision and tried to marry it with an Eastern emphasis on reincarnation. Millenarianism, however, is based on a linear, salvific concept of time while reincarnation, in its Eastern dress at least, is based upon a cyclical view of time. According to reincarnation thinking, one can progress within successive lives, due to a good previous life, but salvation itself is to be found only by escaping history. It is only in certain Japanese Buddhist groups where there appears a belief in a new age, an age of wisdom and bliss. Indeed, according to Palmer, "Hinduism and Buddhism teach that far from being on the verge of a new and improved world, a world which is evolving from the primitive to the sophisticated, we are actually degenerating and are entering the last and most debased era of this planet's current existence."[3]

CONTEMPORARY INFLUENCES

While Gnosticism and millenarianism are important ideological precursors, the specific historical roots of the New Age movement are to be found elsewhere, most notably in Swedenborgianism, transcendentalism, spiritualism, theosophy, and the American counterculture movement of the 1960s.

Named after a Swedish scientist, later turned minister, Emmanuel Swedenborg, the movement that bears his name is a curious marriage of science and religion. A noted scientist credited with anticipating numerous discoveries including crystallography, Swedenborg believed that he could prove scientifically that the universe had a primarily spiritual structure. Curious and realistic dreams convinced him that he had direct access to the spirit world and he spent the latter part of his life studying the Bible and disseminating his teachings.

Swedenborg's influence on the New Age movement was made indirectly through the influence of his thought on transcendentalism, spiritualism, and theosophy. Transcendentalism was best expressed by Ralph Waldo Emerson's book entitled *Nature* published in 1830. A uniquely American form of the Romantic reaction against the rationalism of the Enlightenment, the movement emphasized the importance of feelings, the individual's ability to intuit religious truth, the fluid nature of revelation, God as Spirit, and the importance of incorporating Hindu and Buddhist beliefs into Christian thought. The spread of the movement was due mainly to the efforts of young Unitarian ministers who revolted against the rationalistic creed of Calvinism that was popular in New England at that time.

While Swedenborg expressly rejected spiritualism, nonetheless, the belief in communication with the dead could not be silenced. This belief in communication with the spirits of the dead by the living is as old as the belief in human immortality. If human beings are spiritual entities then we cannot die; after the death of the body our spirits are alive in another dimension, so theoretically, it should be possible to communicate with the spirits. Spiritualism is fueled by this logic and driven by the desperation of people to make contact with loved ones who have died.

The modern roots of spiritualism date from the occult experiences of the Fox family in 1848. Two teenage girls, Margaretta and Katie Fox, reported "rappings" in their home, which they interpreted as messages from a peddler who had died in their house. When Margaretta and Katie Fox clapped their hands, the rapping noises answered back. Working out a simple yes-no code, Margaretta and Katie figured out that the spirit making the noises was a peddler named Charles Rosa who had been murdered by a former occupant of the house named John Bell. Digging in the basement of the house yielded only a few human teeth, hair and some bones, but that was enough to capture the attention of the press and the movement spread quickly.

It is theosophy, however, which has had the greatest impact upon the New Age movement. In its wider meaning, the word refers to any philosophical or theological system based on direct and immediate contact with the divine. In its strict sense, it is commonly applied to a movement started by Madame Blavatsky, a Russian adventurer who, along with H. S. Olcott, founded a Theosophical Society in New York in 1875. Blavatsky and Olcott put together a curious and intriguing blend of ideas supposedly based on Indian sacred books such as the *Upanishads* and the *Sutras* as well as on teachings from Indian Mahatmas. These included a belief in reincarnation, spiritual evolutionism, the deity of the individual as a radiation of the deity of God, the universal truth of all religions, the discovery of esoteric truth in various sacred writings, and the essential oneness of all.

Theosophical teachings were transmitted to the New Age movement by Alice A. Bailey who is credited with having introduced the term "New Age."[4] A prolific writer, Bailey's most famous piece is known as the "Great Invocation":

> *From the point of Light within the Mind of God*
> *Let light stream forth into the minds of men.*
> *Let light descend on Earth.*

> *From the point of Love within the heart of God*
> *Let love stream forth into the hearts of men.*
> *May Christ return to Earth.*

From the center where the Will of God is known
Let purpose guide the little wills of men –
The purpose which the Masters know and serve.
Let Light and Love and Power restore the Plan on Earth.

OM OM OM[5]

THE COUNTERCULTURE
MOVEMENT OF THE 1960S

The teachings of theosophy, transcendentalism, and spiritualism would have remained on the sidelines of North America and Europe but for the political upheavals surrounding the birth of the counterculture movement of the 1960s. In the 1960s former authorities and goals were seen to be empty and vacuous. Life in the suburbs, with a house, two kids, a new car, and a family dog could no longer provide a sense of meaning to many young people. The fragmentation of the movement into such diverse groups as the Hare Krishna, the Jesus People, various New Age groups, socialist communes, fundamentalist churches, charismatic groups and the like masked their essential spiritual unity – their attempt to find meaning and purpose in a world which seemed meaningless.

I remember clearly one poster that my wife gave me at the time. It was a picture of Jesus dressed in a simple, homespun robe. He was fast asleep, hunched over on the front pew of a church while the bald-headed minister droned on to a group of parishioners. The message was clear; the institution of the church was an enemy of true faith, not a friend. The institutionalized church was irrelevant to the spiritual quest. Little wonder that one of the main features of the New Age movement stresses self-religiosity and distrust of traditional religious institutions.[6]

Capping the countercultural revolution of the 1960s, according to many New Age scholars, was the self-help and self-acceptance movement associated with Abraham Maslow. By focusing on the positive as well as the inner person, Maslow helped prepare the way for ready acceptance of many New Age teachings.

Again, there is a curious tension in the New Age Movement between its social vision of a transformed world and its emphasis upon self-fulfillment. Like the tensions between reincarnation and the concept of a new age, as well as the tension between affirmation of the flesh and a belief in a primarily spiritual world, the contrast between societal transformation and inner self-fulfillment serves both as an attractive feature of the New Age movement and the beginning of this movement's possible future collapse.

Although I do not feel that it is fair to dismiss the New Age movement as simply narcissism applied to spirituality, nevertheless, New Agers must watch out for the dangers of a faith which dwells on personal transformation without demanding a concomitant sacrifice for the greater society. As Katherine Smalley, a former television producer puts it, "in the 18 years I spent making films, the real contemporary saints were the ones who laid their lives on the line every day." In contrast, Smalley dismisses much of the New Age movement as "spiritual thrill seeking."[7]

The emphasis on self-fulfillment, however, is not quite as simple as Smalley and other critics would have one believe. It may well be that the emphasis on self-fulfillment is not simply the "me" generation catching religion, but something deeper and more profound. It may be that this attention to spiritual fulfillment is a beginning step in a renewed spiritual growth that reaches out to transform society for the better. As the baby boomers of the 1950s face their own mortality, long-denied questions about life and self come bubbling to the surface, a necessary first step in the spiritual pilgrimage. But, it must be remembered, this is a first step – one which should lead to others.

THE FAILURE OF THE INSTITUTIONAL CHURCH

The crisis of meaning and the turn away from institutional toward individual religiosity are the two most prominent reasons why the New Age has become popular. There is also a third reason – the failure of the institutional church to provide a sense of meaning and purpose, to touch the spirit of the individual with the spirit of the divine. Ted Peters, an American theologian and student of the New Age movement, puts his finger on this reason when he states,

The pastors and priests leading our parishes are for the most part poorly equipped to minister in the new age atmosphere. A number of contradictory reasons for this exist. Although our clergy are allegedly religious leaders, they do not necessarily have a high opinion of religion. Although they preach and teach about things spiritual, they may be quite unsure what spiritual means. Although they lead prayer during worship service every Sunday, they may find little personal fulfillment in private prayer during the week. Although the religious symbols of the Christian religion speak of heaven and hell and things cosmic, many of our pastors have already consigned such things to outdated mythology and superstition. They believe religion is existential and ethical, not metaphysical and cosmological.[8]

In effect, too many clergy have taken Kant's teaching that the divine can only be reached through the sense of "oughtness" and have reduced religion to doing the right thing. But religion, although it includes ethical action, can never be merely doing what is right; it must also encompass a relationship with the ground of our being, with the divine. The horizontal dimension of faith, the ethical, must always be accompanied by the vertical dimension, the mystical. The loss of this vertical dimension in institutional religion is one essential reason why the New Age has become popular.

MAIN TENETS OF THE NEW AGE

At the risk of oversimplification, the main beliefs of the New Age movement are:

1. The conviction that a New Era/Age has dawned that will bring about massive social, political, cultural and economic changes which, in spite of short-term dislocation, will result in long-term benefits. In opposition to the previous age that stressed the importance of the material, this new age will focus on the priority of the spiritual.

2. The priority of the individual in determining their religious faith, along with a distrust of the traditional religious institutions. Also, a belief that the divine resides within each individual and that getting in contact with the divine can empower us.

3. A commitment to monism (all is one) and holism (all is interconnected) which undergirds such practices as channeling and the belief in re-incarnation.
4. A sense of partnership with followers of neo-pagan religions such as Wicca, practitioners of alternative medicine or native/aboriginal spiri-tuality, deep ecologists, and popularizers of the new physics.
5. An attack on the patriarchal nature of Western society and an elevation of the intuitive religious insight of women, coupled with support for the concept of a pre-Jewish era in which worship of the Goddess prevailed.

A NEW AGE

The most important New Age tenet is the belief that we are entering a new age. The best publicized form of this vision (as indicated by the title of Marilyn Ferguson's book *The Aquarian Conspiracy*)[9] claims that the Age of Pisces (the Fish) which was dominated by the institution of the Christian church is being displaced by the Age of Aquarius (the Water Bearer). An-other New Age leader, José Argüelles, uses an old Mayan calendar rather than the Zodiac symbols to advance the claim that we are entering a new age, which he calls the age of Harmonic Convergence.[10] Riane Eisler makes a much more penetrating social analysis and argues that we have moved from a "gylanic" age in which men and women were equals, through an age dominated by patriarchy, to a new age where the perspectives of the first age are recaptured. This new age, she states,

...will be a world where the minds of children – both girls and boys – will no longer be fettered. It will be a world where limitation and fear will no longer be systematically taught us through myths about how inevitably evil and perverse we humans are. In this world, children will not be taught epics about men who are honored for being violent or fairy tales about children who are lost in fright-ening woods where women are malevolent witches. They will be taught new myths, epics and stories in which human beings are good; men are peaceful; and the power of creativity and love – symbolized by the sacred Chalice, the holy vessel of life – is the governing principle.[11]

A more specifically religious expression of this theme of a new or third age can be found in the claims of Emilia Rathbun who, along with her husband, succeeded Henry Sharman as leader of a group known as the Creative Initiative. While this particular group predated the rise of the New Age movement, during the 1970s it advocated many New Age themes. Thus, Emilia claimed that God was calling for a new covenant community, based on a new revelation. Stephen Belber and Martin Cook who have studied this movement note, "although she [Emilia] was unwilling to make the claim in public, in private Emilia was willing to affirm that she was to the covenant of the third age what Jesus was to the second and Moses was to the first."[12]

This emphasis on a new or third age has both drawbacks and attractive features. The main attraction for me rests in the optimistic spirit with which most New Age groups face the future. While there are some who call for tribulations in the future, by and large the New Age movement provides a refreshing counterbalance to the pessimistic spirit of premillenialism, which after the First World War dominated conservative Protestants. Instead of the turn of the millenium producing the four horsemen of the Apocalypse as in premillenialism, the new age is to usher in a beatific vision of harmony and peace. Eisler's comments, in particular, strike me as nothing more than a politically correct version of the biblical prophet Isaiah's views. In Isaiah's vision of the future of the Israelites, interpreted by the Christian community as a reference to Jesus, Isaiah writes, "the wolf shall live with the lamb, the leopard shall lie down with the kid, the calf and the lion and the fatling together, and a little child shall lead them" (Isaiah 11:6 NRSV).

Another attractive feature, one that as I already mentioned conflicts with support for the theory of reincarnation, is the emphasis upon salvation within human history. Although New Age adherents are usually very vague about when this is going to happen, it is hard to quarrel with the vision of a world where man is no longer at odds with woman, where the human creature is at peace with nature, and where the divine speaks directly to the heart of each person – a vision that is possible in the here and now, not just in the life to come. Likewise the Christian faith, which I hold as my own, has at its center the concept of the atonement which, whatever else it may mean, underlines the activity of the divine within time and history.

In this regard, the New Age movement, often criticized for self-centeredness and escapism, actually forms a corrective to the tendency in some Christian circles to focus on salvation in the future, on the other side of death, rather than on salvation in the here and now. It serves to remind Christians that God has become one with creation through the *at-one-ment*, and its implications. Moreover, the social implications that are part of this focus on the present could help counteract the extreme individualism of Western culture. The New Age is not conceived as a private event but as a social transformation; one that will bring peace and harmony to all.

The drawback of this new/third age vision is one common to all revolutionary movements. It collapses the past and present into the immediate future. Time becomes now, a point, rather than a line. Little wonder that many revolutionary movements end up perpetuating the mistakes of the past! Moreover, the fullness of activity of the divine, safeguarded by that awkward yet important Christian doctrine of the Trinity, is also reduced. The emphasis on the transcendence of the divine or on the tragic within life, which caused the crucifixion of the Christ, is forgotten. The divine becomes a one-dimensional being in spite of all the rich spiritual language one encounters in New Age literature.

SELF-RELIGIOSITY AND A DISTRUST OF INSTITUTIONAL RELIGION

The second key theme that emerges in the New Age movement, as already noted, is the emphasis on the individual in determining his/her religiosity coupled with a distrust of religious institutions. A direct outgrowth of the counterculture movement of the 1960s, the distrust of old-line religious institutions is a common feature in the messages of the various New Age channelers.[13] In part, this is because the channelers compete with the religious institutions for financial support. It would be wrong to underestimate the financial factors involved in the growth of the New Age movement. As Palmer notes, "to a certain extent, the phenomenon of the New Age is a publishing phenomenon, in part created, certainly sustained and possibly even at times invented by a publishing world that has sensed the spiritual and religious hunger, and has set out to make money feeding it."[14] Of course, as Palmer also notes, the Protestant Reformation was also a publishing phenomenon and the tracts and writings of the Reformers were bestsellers. The difference that Palmer fails to mention is that sometimes the publishers of those tracts received not monetary reward as do the publishers of New Age material, but persecution and even martyrdom!

Desire for individual control over one's spiritual journey, however, cannot simply be attributed to publishing houses and New Age leaders out to make money. In the world of medical treatment, politics, education, and religion, individuals are rising up and demanding a say. This explains the significantly high levels of religious interest in North America as evidenced by various polls.[15] Or as Lyle Schaller, a popular church-growth analyst, quips, "You can still find a high degree of receptivity to institutional control in three places: retirement homes, nursing homes and cemeteries."[16]

Even within that group of people who remain loyal to traditional religious institutions, one finds a demand for control asserted through what has been called cafeteria-style church affiliation. For example, a family may attend their local Methodist church for Sunday morning worship. The children may be enrolled in the Baptist youth group, while during the summer the mother may help at the local Presbyterian daily vacation Bible school. Moreover, church-hopping has become a common practice, as individuals will move from church to church following the action.

This self-religiosity is coupled in New Age teaching with a stress on the divinity of the individual. Again, one can see this as reaction arising out of the counterculture movement. The explosion of suburbia following the Second World War in the United States and Canada along with the proliferation of cheap, mass-produced goods, resulted in the acquisition of material possessions on a scale unheard of in previous generations. The negative side of this explosion of material wealth was the bureaucratization of society. No longer did one shop at the local store where one was known and one's family history could be recounted; now one shopped at a department store where plastic cards with red or black numbers printed on them expedited the buying process. The loss of individuality and the reduction of life to the mere accumulation of material goods created a vacuum of personal worth and meaning.

The New Age message that you are divine, a god or goddess, restores the sense of individual worth and meaning. So what if you live in a depersonalized, mechanistic world. The true you is a spiritual being capable of contact with the spirit world. Elizabeth Hillstrom criticizes New Age channelers because they provide similar messages, thus, to her mind, exposing them as frauds. But then she summarizes these messages which stress the worth of the individual, the importance of trust, and an optimistic future.[17] Even if wrong, it is hard to see how such messages hurt individuals and easy to see how they might help.

The price of this stress on the divinity of the individual and self-religiosity, however, must also be mentioned. If we are all gods, then how do we account for the wrong that we do or think or feel? We are ambiguous creatures capable of great love and equally great hate. The theory of the essential divinity of the individual leaves no room for the tragic. Of course, one response would be that the tragic – what Christians call sin and Marxists call alienation – is ephemeral, in essence non-existent, a figment of the imagination. Such a response fails to do justice to life where we hurt and are hurt, and *we know it*. It may well be that the emphasis of the Christian New Age writer, Matthew Fox, on original blessing needs to be stated in face of Western Christianity's focus on original sin. But surely any realistic view must come to terms with both blessing and sin.

Moreover, the private religiosity of the New Age movement further erodes the sense of community that is so desperately needed today. Hanengraaff takes this criticism even further when he concludes his book on the New Age movement by arguing that the privatization of faith in the New Age results in a dissipation of mystery as each private individual is made "the center of his or her symbolic world."[18] The philosopher Alfred North Whitehead is credited with saying that religion is what one does in one's solitude, but this is not totally true. Both solitude and social interaction are needed in a symbiotic mix, at least according to the tenets of the Christian faith.

MONISM AND HOLISM

A third feature of the New Age movement is monism and holism coupled with an evolutionary view of human history, based mainly on the thought of the Roman Catholic theologian Teilhard de Chardin. Often such monism and holism has led to the attack that New Age proponents are pantheistic, worshipping sticks and stones as if they were God. This is patently untrue; a better description would be that most New Age participants are panentheistic. They see God as being greater than the created order, but as being immersed in that creation as well – spirit interpenetrates matter and cannot be separated from it.

This holism is a welcome counterbalance to the specialization that has reigned in Western society as the modern scientific world view predominated. The individual, the world, and society were seen to be made up of smaller bits that could be taken apart and then put back together at will. Reality lay in the parts not in the whole. The New Age movement can be seen as a Platonic revolt against the Aristotelianism which dominates our thinking and supports the concept that reality lies in the individual parts rather than the unified whole. The humorous saying that an academic is one who knows more and more about less and less until finally they know everything about nothing is an example of this protest. In New Age thought, the whole is greater than the sum of its parts.

A commitment to monism and holism accounts for the sense of partnership and unity which exists between New Age devotees, practitioners of alternative medicine, deep ecologists, popularizers of the new physics, and neo-pagans. Important differences exist between these groups, but at heart there is agreement. They agree that society needs to change and that the reign of rationalism and traditional science has ignored rich spiritual truths which are necessary if humankind is to continue. Indeed, so seamless does the partnership between these groups appear to the outsider that Henry Gordon, a Fellow of the Committee for the Scientific Investigation of Claims of the Paranormal, a skeptical rationalist think-tank which tracks the New Age movement, goes so far as to claim that "of all the potential harmful effects that can be laid at the doorstep of the New Age movement, I would say that the current promotion of alternative medical practices is the most damaging."[19] Michael Cole, a conservative Christian critic, makes much the same assumptions about the ecological movement, claiming, "whether it be the moderate view voiced by the likes of Jonathon Porritt, the Earth mysticism of the Gaia Hypothesis, or the strange world of Devas, nature spirits and Pan experienced at Findhorn, one thing is clear. All see conservation as a religious matter. It is one of the most powerful influences leading people into the New Age movement."[20]

The critics of the New Age movement, however, overstate the connections between these groups. Many people visit chiropractors with no interest whatsoever in the New Age, while countless numbers are rightly concerned about the environment who would scoff at the Gaian hypothesis and have little to do with nature mysticism. What ties the New Age movement together with these other groups is the fact that Western society has arrived at the edge of a crisis of spirit. New ways of looking at life have arisen in response. It is the current milieu, often called post-modernism, which binds alternative medicine, the impact of the new physics, the New Age movement, and deep ecology together, rather than institutional links. It is a common search brought about by common problems.

MYSTICISM AND THE EROTIC

Another positive aspect of New Age teaching is that it helps put the erotic back into spirituality. Indeed, to the outside observer many New Age practices are immoral because they are erotic. Contrast, for example, a typical communion or Eucharist service with a Wicca ceremony. In the communion service, everyone sits in their Sunday best on a hard wooden pew. In some churches one does not even have to get up to get the bread or drink from the cup, since they are delivered to your seat – sanitized square pieces of bread, and wine in small, safe glasses. In a typical Wiccan celebration, body movements, touch, taste, nudity, smell and sound are interwoven in a rich, sensuous mix. And yet, Christian spirituality at its best has always had an erotic element as the mystics of the past have known and expressed. In this regard, I cite the Jesuit poet Gerard Manley Hopkins who in the poem "God's Grandeur" protests against some of the effects of the Industrial Revolution and writes,

> *The world is charged with the grandeur of God.*
> *It will flame out, like shining from shook foil;*
> *It gathers to a greatness, like the ooze of oil*
> *Crushed. Why do men then not reck his rod?*
> *Generations have trod, have trod, have trod;*
> *And all is seared with trade, bleared, smeared with toil;*
> *And wears man's smudge and share's man's smell: the soil*
> *Is bare now, nor can foot feel, being shod.*
>
> *And for all this, nature is never spent;*
> *There lives the dearest freshness deep down things;*
> *And though the last lights off the black West went*
> *Oh, morning, at the brown brink eastward, springs –*
> *Because the Holy Ghost over the bent*
> *World broods with warm breast and with ah! bright wings.*[21]

This mystical oneness, however, must also recognize "the other" and, in recognizing the other recognize ethical limitations designed to protect the other. Erotic spirituality can sometimes treat the other as simply an extension of the self and cross boundaries, which should not be crossed on this side of heaven. Moreover, absorption into "the one" has never been a source of consolation to me. I long to feel that sense of surrender to the divine, that ecstasy of being lost to self, but I also long to remember self and to recognize the other. In a sense, I want my cake and I want to eat it too. Yet I think that the paradox of plurality and unity cannot be dissolved without doing injustice to life. One cannot surrender to the other unless there is another to surrender to. The very thrust of oneness demands distinction as well as unification.

THE RELIGIOUS INSIGHT OF WOMEN

A final essential feature of the New Age movement is the emphasis on the intuitive religious insight of women. The New Age has benefited from the religious vision of women and has reciprocated by calling for gender equity and the displacement of male religious symbols by practices using female ones. This is most evident in the Wicca faith, which although it is more properly categorized as part of neo-Paganism, can also be seen as part of the New Age movement. In Wicca, for example, the priestess holds a higher office than her male counterpart, and worship of the Goddess is prominent.

REACTIONS TO THE NEW AGE

Reactions to the New Age movement have been varied. The press and the publishing industry almost uncritically have bought into the hype of the movement. One would think from magazine covers that the New Age movement was the dominant religious movement in North America and parts of Europe. Yet statistics of New Age involvement (even including neo-Pagan involvement) show that compared to traditional religious institutions only a small number of people actively engage in it. Indeed, Palmer insists, "I simply do not see evidence of a New Age Movement. I do see evidence of a small and eclectic group of people who bunch together, for a vastly different array of reasons, under the title New Age."[22] Nonetheless, the movement is growing. A 1989 Gallup Poll in the United States revealed that over 12,000,000 people were "New Age inclined."[23]

The movement's growth is one reason it has attracted its share of attention. On the critical side is a curious mixture of conservative Christians and skeptical rationalists. A typical conservative Protestant response can be found in the small paperback book entitled *What Is the New Age?* written by four British Christian leaders. They define the New Age movement as "a man-centered religion seeking 'religious experience' largely for its own sake through manipulations of the human mind, by oriental religions and occult techniques." The authors conclude that since the New Age movement is infiltrating Christian churches it is "a preparation for and will be completed by the return of the New Age, interfaith Christ. This Christ is not Jesus Christ but what the New Testament describes as the Antichrist."[24]

In contrast, a typical skeptical rationalist would be appalled by the conservative Protestant emphasis on the New Age movement as the harbinger of the antichrist. Instead, the skeptical rationalist would throw cold water on the New Age movement by mocking the extremes of the movement and by calling for clear-headed rational analysis. A pointed example of such a critique can be found in the book *New Age Thinking* where the author claims that the New Age movement is simply an immature psychological attempt to restore "the perceptual and emotional omnipotence that characterizes the phenomenology of the neonate."[25]

A more balanced approach can be found in authors such as Ted Peters and Martin Palmer, both of whom recognize good elements within the New Age movement as well as dangers. Thus, Peters appreciates the stress on spirituality which he finds in the New Age movement but is distressed by its deviation from the Lutheran concept of salvation by grace alone and the resulting salvation by works. For his part, Palmer is attracted to the emphasis upon the goodness of nature but is concerned about the syncretism of the New Age movement which in his opinion results in the trivialization of serious religious beliefs. Included with this more moderate group are popular authors such as Sam Keen and Thomas Moore, to name two, who have incorporated aspects of New Age thought into their Christian faith and who are often listed as proponents of the movement.

CONCLUSIONS

Clearly, I have sympathies with certain elements within the New Age movement. While I balk at many practices and find much that smacks of Hollywood, I could make the same criticism of many elements of Christianity. I cannot easily differentiate between the television evangelists and the television channelers. To me they seem like mirror images of each other, too often preying on the hopes and dreams of the naïve and the desperate. I would echo Christopher Lasch's comments when he writes,

The intuition underlying New Age movements, the bedrock feeling, hard to put into words but scarcely inchoate or confused, deserves better than ridicule: that mankind has lost the collective knowledge of how to live with dignity and grace; that this knowledge includes a respect not just for nature but for the nurturant activities our society holds in such low esteem; and that man's future depends on a renewal of prematurely discarded traditions of thought and practice.[26]

My disagreement with the movement is twofold. The first is already alluded to in the comments of Christopher Lasch. Christianity is not a unified, monolithic system of thought and practice. There are discarded traditions in Christianity which have as much power to heal the spiritual sickness of the present as anything I can find in the New Age movement. Moreover, these rich veins are situated within a 2,000-year-old mine which in its Jewish roots goes back even further. The New Age movement, in my opinion, prematurely discards traditions and practices from Christianity.

My second disagreement is that while it is interesting and beneficial, the New Age movement is only a necessary corrective to the overly rationalistic and institutional form of Christianity we have inherited in the West; it is not yet the answer. For example, rather than a holistic reaction to the specialization of knowledge and life, we need a combination of opposing elements held in creative tension with each other. The male, for instance, cannot swallow up the female nor the female the male. Both are needed to discover as full an understanding of life and of the divine as is possible in this finite world of ours.

In the end, I resonate with the claim that the opposite of a fact is a falsehood, but that the opposite of one profound truth may very well be another profound truth. Thus, New Age attacks on Christianity and Christian attacks on the New Age are not as productive, in my view, as finding the paradoxical truths present in both.

5

NEAR-DEATH EXPERIENCES

By the time I got to know Gerry, he was already retired. I wished I had known him earlier. Gerry and his wife had been married just before the start of World War II. When Gerry came home from the war, he and his wife decided they would like to start a family. However, for one reason or another, they were unable to conceive and by the time they decided to look into adoption, they felt that they were too old to start a family. Disappointed, Gerry decided that he would compensate for this lack of family by saving for his retirement. With the money that would have gone toward raising children, he and his wife would travel; they would celebrate and would have fun. On the eve of his retirement, though, when his travel plans were to be initiated, Gerry suffered a massive heart attack. He survived, but his dreams of travel burst. His doctors cautioned him to be very careful and not to exert himself. Although polite, Gerry became a bitter man at heart, bitter toward life and bitter toward God. He resented his wife's involvement in the church I was pastor of and he sought to limit that involvement as much as he could.

In time, Gerry had another heart attack and ended up in the hospital. I would visit him, although probably not as much as I should have. Although he tried hard not to show it, I represented God and he was mad at God and therefore at me. One day his wife phoned me: the doctors had just been in to visit Gerry and they told him that there was nothing more that they could do for him; they told him that he was going to die. She asked if I could visit him to try to give him some hope. I said I would, inwardly cursing the day I decided to become a minister. I didn't want to visit a person who was dying and I certainly did not want to face Gerry's bitter anger.

I drove slowly to the hospital and took my time walking to Gerry's room, but when I entered, something was different. Gerry was lying in bed with an unearthly smile on his face. I knew something strange had taken place. I did not even say hello; I just stood there with my mouth open and blurted out, "Gerry, what has happened?" He smiled and told me that he had had a vision. It was a vision of a circle of lights. Outside the lights, all was darkness, while inside he saw blue skies and green hills. And then he heard a voice, which he took to be the voice of God, say to him, "GERRY, YOU ARE GOING TO DIE."

"Were you frightened?" I asked. "Were you angry?"

"No," he replied. "I felt a wonderful sense of peace." Then he smiled a smile that seemed as if it came from heaven itself.

Goosebumps covered my arms, the hair on the back of neck stood up. I felt as if I should take off my shoes. I felt as if I was standing on holy ground. Two weeks later Gerry died. I had the privilege of being with him. As he died, in my mind's eye, I saw him limp through the circle of lights to the green hills in the distance.

Near-death experiences have become a hot topic recently, the subject of academic studies, popular books and television shows. Strictly speaking, Gerry's experience, which occurred when I was a young pastor in Ontario, was not a near-death experience, but a pre-death vision. Whatever it was, it gave Gerry and his wife and all who knew him and heard the story a wonderful sense of comfort. We all knew that Gerry had been mad at God and we all breathed a sigh of relief that he had finally, just in time, made his peace.

Some time later, I was involved in another event, again not a near-death experience, but rather an apparition. Jane (not her real name) was a down-to-earth practical woman, exceedingly gracious and very dignified. Her husband was a doctor, well known in the community. They had a good marriage and parented four children. Each year, Jane's husband would buy her a dinner bell for Christmas. She kept the dinner bells, mostly made out of china, behind a glass cabinet. Over the years of their married life together, she had accumulated a large and interesting collection.

While I was their pastor, Jane's husband died of a heart attack. With her characteristic dignity and grace, she carried on, keeping her grief to herself, since, in many ways, Jane was an extremely private person. Her husband died in the spring and, as Christmas grew closer, I made a point of contacting her because it is often at Christmas that the loss of a loved one strikes the hardest. One day, after some initial embarrassment and my promise not to tell anyone, Jane confided to me a story.

She was on her way to visit a friend. When she got out of her car and was starting to go up the walkway to her friend's house, a young girl appeared, as if out of nowhere. The girl was dressed in a pink snowsuit. She handed Jane a gift, all wrapped up for Christmas. "Oh, no dear," Jane said, not knowing who the young girl was, "you keep it." The young girl insisted and finally, so as not to hurt her, Jane took the gift. She looked down at it for a moment and then turned to thank the young girl who suddenly had disappeared. Later, when alone at her home, Jane opened the gift. Inside was a dinner bell!

Ever since our species appeared on this earth, it seems, most of us have believed in some sort of afterlife. From tribal societies to modern ones, this belief has persisted, evidenced by deliberate burial practices found by archaeologists in all ancient societies and groups. As researcher Cottie Burland concludes, "a belief in an afterlife seems to be endemic among the human race."[1]

Even the early Jewish community, whose focus was very much this-worldly and whose hope of immortality rested in the historical continuance of the Jewish people rather than in an individual afterlife, believed in a place called *Sheol* – a sort of shadowy afterlife where everyone, good or bad, ended up.

It is really only since the middle of the 19th century in the West that a large number of people, including much of the intelligentsia, have abandoned a belief in an afterlife. And, with the rise of modern science and the retreat of religion to ethics, the belief that this life is all that there is has gained momentum. I will never forget visiting a woman whose baby had died from Sudden Infant Death Syndrome (SIDS). The death of her baby had convinced her that a good and loving God could not possibly exist. Not knowing what to say, and wanting to comfort her, I asked, "And where is your baby now?" I hoped to follow up with the assurance that her baby was with God, but she would have none of it. "He's dead," she replied, "in the ground, rotted away."

It seems, however, as if we cannot live long without the hope of some afterlife, either because we are emotionally weak, as skeptics claim, or because the afterlife is real, as believers claim. As evidence of this, witness the recent emphasis on the reality of the afterlife which has arisen in Western society. Interestingly enough, this not arisen from within the Christ church, but from outside the church, from within the field of science itself.

THE HISTORY AND STUDY OF NDES

Most date the renewal of this emphasis back to 1975 and the publication of the book *Life after Life* by a physician, Raymond Moody, who is credited with coining the phrase – near-death experience (NDE).

The genesis, however, can really be traced back to an event that happened on January 12, 1924, in Dublin, Ireland. Lady Barrett was a physician who specialized in obstetrical surgery. She had been called to deliver the child of a woman named Doris. The baby was born safely, but the mother was dying. Lady Barrett described it like this.

Suddenly she looked eagerly toward one part of the room, a radiant smile illuminated her whole countenance. "Oh lovely, lovely," she said. I asked, "What is lovely?" "What I see," she replied in low, intense tones. "What do you see?" "Lovely brightness – wonderful beings." Then – seeming to focus her attention more intently on one place for a moment – she exclaimed, almost with a kind of joyous cry, "Why it's Father! Oh, he's so glad I'm coming; he is so glad."[2]

Doris's vision so impressed Lady Barrett that she gathered various such stories and published them in 1926 in a book entitled *Death-Bed Visions*.[3] It wasn't until Elisabeth Kübler-Ross began her work on the process of dying, however, in the 1960s and 1970s, that much scientific attention was paid to these deathbed visions. And it was, shortly after that, as already mentioned, that Raymond Moody published his popular book *Life after Life*,[4] a collection of accounts from people who had been close to death and then medically resuscitated. Most had suffered cardiac arrests, although not all. Out of their stories, Moody discerned a common experience, with several typical features which he called a near-death experience.

Soon, others began to take an interest in the commonalities that Moody discovered and attempted to advance his findings. In 1977, a psychologist, Kenneth Ring, became interested in NDEs. Later, in 1978, Ring founded the International Association for Near-Death Studies to better understand and legitimize this phenomenon.

Another doctor, Melvin Morse, has pioneered the study of NDEs in children. An experience treating a young girl named Katie, who almost drowned, gripped Morse's imagination. Morse later described the incident.

In 1982 as a pediatric resident, I was examining one of my patients, a little girl named Katie, who had almost drowned in a community pool in Idaho. Even without her near-death experience Katie was a remarkable story. She was documented as not having a pulse for nineteen minutes. When I first saw her, her pupils were fixed and dilated, meaning that irreversible brain damage had most likely occurred.

I worked hard on her anyway, although in my heart I didn't think she would survive...

Three days later she made a full recovery.

One afternoon I casually asked her what she remembered about being in the pool... the answer she gave me wasn't anything like I had expected: "Do you mean when I saw the Heavenly Father?"

After Morse's initial shock, Katie continued with her story. She told him of the treatment the doctors had performed on her during those three days, as well as of a trip through a long, dark tunnel and a visit with an "angel" who calmed Katie down and took her to look in on her brothers and her mother during those three days. At a loss for words, Morse asked Katie what it was like up there. She replied, "You'll see, Dr. Morse. Heaven can be fun."[5]

More recently, researchers have been examining the question of why Morse, Moody, Ring, and others failed to report many negative near-death experiences. Maurice Rawlings has been the main figure in the investigation of *negative* experiences which, since Rawlings is a committed evangelical Christian, has prompted accusations of bias by other authors more inclined to support some form of New Age faith. As Cherie Sutherland notes, "In 1978 Rawlings presented the thesis that hellish NDEs are simply repressed. Arguing as he does, however, from a 'born-again' Christian perspective with the clear agenda of proving to readers the existence of hell and therefore the need to be 'saved,' *his presentation is questionable*."[6] Rawling's findings, however, are slowly being supplemented by other researchers. Margot Grey, a founder of the International Association for Near-Death Studies in the United Kingdom, observes, "About an eighth of my respondents reported experiences that were hell-like; this corresponds more closely with the findings of Maurice Rawlings and George Gallup rather than with those of Kenneth Ring and Michael Sabom."[7] Clearly, more study needs to be done on negative near-death experiences as well as on cross-cultural variations, which is another area of disagreement between researchers.

In fact, the study of NDEs is in its infancy. In spite of claims to the contrary, the scientific community as a whole has dismissed the experience or else has treated it as a type of hallucination brought about by chemicals released by the brain during moments of crisis or by a lack of oxygen. Historical studies also are scarce. Several popular authors point out interesting parallels in the writings of Plato and in the experiences of the apostle Paul. In the tenth book of Plato's *Republic*, he relates the tale of a man named Er who was killed in battle. Twelve days after his supposed death, Er came back to life. Plato narrates the story of Er's experience.

...when the soul had left his body, he journeyed with many others until they came to a marvelous place, where there were two openings side by side in the earth, and opposite them two others in the sky above. Between them sat Judges who, after each sentence given, bade the just take the way to the right upward through the sky, first binding on them in front tokens signifying the judgement passed upon them. The unjust were commanded to take the downward road to the left, and these bore evidence of all their deeds fastened on their backs.[8]

Another famous historical NDE is supposed to have happened to the apostle Paul, based on his comments (likely autobiographical) about a man, "who fourteen years ago was caught up to the third heaven – whether in the body or out of the body I do not know; God knows. And I do know that such a person... was caught up into Paradise and heard things that are not to be told, that no mortal is permitted to repeat" (2 Corinthians 12:2–4).

Closer to our own time, the famous thinker Carl Jung reported a near-death experience which happened to him in 1944 following a heart attack. Jung writes:

In a state of unconsciousness I experienced deliriums and visions which must have begun when I hung on the edge of death and was given oxygen and camphor injections. The images were so tremendous that I myself concluded that I was close to death. My nurse afterward told me, "It was as if you were surrounded by a bright glow." That was a phenomenon she had sometimes observed in the dying, she added. I had reached the outermost limit, and do not know whether I was in a dream or in ecstasy. At any rate, extremely strange things began to happen to me.[9]

Jung's account covers several pages and begins with him zooming up through the sky until he had a bird-like view of the entire earth. At this point, he notices a nearby meteor in the shape of an Indian temple, which he enters. The vision ends when an image of Jung's doctor floats up to him from the direction of Europe, "to tell me that there was a protest against my going away. I had no right to leave the earth and must return. The moment I heard that, the vision ceased."[10]

While the accounts found in the writings of Plato, the Bible, and Jung are interesting and are used by NDE advocates to buttress the truth of their position, they are problematic. For one thing, there are only a few of them. This, I suppose, could be explained by the medical advances of the 20th century which have enabled doctors to resuscitate many more people who would previously have been left to die. However, the more telling criticism is that even the few NDEs that exist do not conform to the stereotypical pattern defined by Moody. Plato's account, for instance, sounds much more like the parable of Jesus found in Matthew chapter 25, about the division of humanity into sheep and goats, than it does a modern NDE. Obviously, Plato was using a well-known myth to teach that we will pay the price for our wrongdoing in the future and therefore we should seek to live ethically in the present. Paul's account, while intriguing, is not full enough to place much weight on it, while Jung's account differs markedly from those examined by Moody and others.

CHARACTERISTICS OF
THE NEAR-DEATH EXPERIENCE

So what *are* the features of a typical near-death experience? In book after book, Moody's account is repeated with minor variations.[11] The first component is a sense of being dead. Initially, "experiencers," as they are called (an interesting word!) are afraid and disoriented and sense that they are dead. Once this realization touches them, the initial disorientation passes and a sense of peace prevails. This is usually followed by an out-of-body experience (OBE) – NDErs it seems cannot get enough of acronyms – in which the person feels themselves separating from their body and then floating above their body, looking down at it. Sometimes the person tarries, moving through walls at will and observing the activity of the medical staff trying to resuscitate them.

In time, the person journeys through a long, dark tunnel that leads to a bright light at the other end. What the person sees at the end of this tunnel has been described in various terms ranging from urban to pastoral. Most people see pastoral settings, as was the case with Gerry, some see more urban images such as temples and other elaborate buildings.

Suddenly, the person becomes aware of "beings of light" which may be religious figures, dead relatives, or unknown apparitions. These beings of light provoke a feeling of indescribable love, which is heightened by contact with the ultimate being of light, usually interpreted to be God. The task of this being of light is to initiate the process of a life review whereby the "experiencer" sees the events of their life pass before their eyes. Sometimes this is intended to help the person see the wrong that they have done or the pain they have inflicted on others. Most often, the life review provides the person with answers to why the different events of their life unfolded as they did. The life review is the final portion of the NDE after which the person is told that they have to return to this world. However, the experience is such a profound one that most people experience various aftereffects that transform their personality, almost overwhelmingly for the better.

Tom Harpur, the well-known Canadian professor and religious writer, has been so impressed by the common elements within the experiences of people who have come close to death that he has embarked upon a study of what major religious groups throughout the world claim about death and the afterlife. In his opinion, the major world religions, including native spirituality, "say amazingly similar things about life beyond death." He continues, "What's more, as you study them, the similarities to what is being described by those who have had an NDE begin to leap out at you." A specific example, according to Harpur, can be found in the Tibetan Buddhist *Book of the Dead* which contains "startling" parallels to Moody's tunnel of light.[12]

Harpur has also been interested in the aftereffects exhibited by the experiencers. These include:

- a marked decrease in the fear of death paradoxically coupled with a realization that suicide is not a suitable option for dealing with the troubles of life;
- greater compassion and love for other people which often leads to career choices in the helping professions;

- an openness to mystical experiences which leads to claims of various psychic abilities along with a greater emphasis on prayer, meditation, and exercise than is the case with the average person;
- a devaluation of the pursuit of money as the goal of life, replaced by an emphasis on spiritual growth.

While all these aftereffects would be labeled positive, it is clear that the impact can be a negative one, especially when it comes to the person's most intimate relationships. In fact, a common aftereffect is marriage breakup; the experiencer has changed so much that the marriage relationship cannot adjust. Cox-Chapman observes that, "many experiencers find that their relationships cannot absorb the changes near-death experiences bring. Although experiencers say they are more focused on love since their NDE, the divorce rate among experiencers is slightly higher than the national average."[13]

Typical elements of *negative* near-death experiences parallel positive ones, with some significant differences. The initial stages of the experience are similar, but rather than entering into light, peace and joy those who have endured a negative NDE feel themselves entering into a black void where they come in contact with an evil force. Angie Fenimore wrote one of the most interesting book-length accounts of a negative experience. As Angie enters the black void, coming in time to a depressing, hell-like environment, she begins to sense a loss of identity and an accompanying loss of brightness:

I sensed that I wasn't entirely female anymore, I was the same individual that I had been before – my morbid sense of humor, my curiosity, my personality, the way I thought and felt remained, and my awareness of being female was also with me – but my form had been somewhat reduced, made not smaller but less complicated...

Everyone I saw was wearing dirty white robes, Some people's were heavily soiled, while others' just appeared dingy with a few stains. I sensed that the housecoat I had on when I lay down to die had been replaced by dark clothing, possibly a familiar black sweater I had worn often that winter.[14]

The aftereffects of a *positive* near-death experience and of a *negative* near-death experience, however, are not very different. Since many negative experiences occur among those who attempted suicide, the clear message is that suicide is not a healthy option. Nonetheless, few attempted suicides feel that they have been judged in a condescending manner, but rather that they have been loved by God and given the opportunity to be kinder and more loving to others.

The well-known theologian, Hans Küng speculates that negative NDEs may be tied to the cause of death. According to Küng, positive NDEs occur to people dying slowly from cancer while different NDEs might likely occur in people who die quickly.[15] This, though, is speculation. At present, the only common thread concerning negative NDEs is the person's previous life (hence the connection with suicide) as well as the fact that negative NDEs are usually reported *immediately* after the person's resuscitation. The importance of the timing of the report of the negative experience is yet unknown. It may be that people who have had negative NDEs are embarrassed to reveal their experience and so after the initial shock – when they are more susceptible to sharing the event – they clam up. Alternatively, it may simply be a case that over time our memories filter out the bad and magnify the positive.

RESPONSES TO THE
NEAR-DEATH EXPERIENCE

How then should we interpret NDEs? Some claim that a variety of physical or psychological reasons cause these experiences; others claim that the vast majority of NDEs are mystical experiences of a very real afterlife. Those who favor physical or psychological causes have advanced several arguments in support of their thesis. Susan Blackmore, for instance, notes that she personally experienced an "NDE-type experience complete with the tunnel and light, out-of-body travels, expansion and contraction of size, timelessness, a mystical experience and the decision to return" as a result of extreme tiredness coupled with smoking hashish.[16] Her main thesis, however, is that NDEs are caused not by drugs such as hashish or LSD but by the suffocation of the brain. According to Blackmore and others, as the body dies, the brain receives less and less oxygen, which triggers a near-death experience.

Closely allied to the dying-brain syndrome is the theory that as the brain begins to die it releases chemicals which act as natural painkillers and anxiety soothers. It is these endogenously produced chemicals, such as endorphins, that trigger NDEs. Still others argue that the process of dying provokes temporary seizures of the brain and it is these, which give rise to NDEs.

Those who favor psychological explanations often claim that NDEs represent wish fulfillment. The dying person wishes to live forever in a place of light and beauty and so a vision of such a place is produced by the person's psyche. Coupled with this is the suggestion that the out-of-body experience, which is part of the NDE, is really a defense mechanism. The person detaches themself from the dying body and experiences what is known as transient depersonalization. The most intriguing psychological explanation, however, is one that has been proffered by Carl Sagan, the popular science writer. He suggests that near-death experiences stimulate long buried memories of a person's birth which come floating back to the surface. He equates the long tunnel with the birth canal and the bright lights with the stereotypical hospital delivery room.[17]

Advocates of a mystical interpretation (whereby the NDE is seen as a valid glimpse into the afterlife) discount both physical and psychological explanations. Instead, they believe that science has simply not caught up with the near-death experience. As Rosalind Heywood puts it,

I have wondered, when watching the spokes of a wheel revolving too fast to be visible, or when seeing an ultrasonic whistle being blown which was silent to me, whether the possible existence of speeds of movement or vibration, beyond the adult human being's senses or imagination, might have something to do with our lack of concepts for conditions in which discarnate consciousness might exist. A few days after reading about tachyon hunters [a tachyon is a hypothetical elementary particle that is able to travel faster than light] I was told about a little girl whose mother had drowned. The day after the disaster her father shut himself up alone with his grief, but she insisted on seeing him. She had some news. "I've been talking to Mummy," she said, "and she says, 'Not to worry.' She's still with us, but going faster so that we can't see her."[18]

Melvin Morse, the doctor who studies in NDEs, hypothesizes that Western science cannot come to terms with the near-death experience due to the predominance of reason within Western civilization and a consequent atrophying of the use of that part of the brain which can understand and communicate with the divine. He writes:

After examining thousands of case studies, and even after having had a death-related vision of my own, I can say without a doubt that the brain both creates visionary experiences and detects them. There seems to be a huge area of the brain that is devoted to having just such experiences. Just as we have a region of our brain devoted to speech and one that helps us regain our balance when we trip and almost fall, we have an area that is devoted to communication with the mystical. It functions as a sort of sixth sense. In short, it is the "God sensor." [19]

Morse's assertion makes sense of two seemingly contradictory stances taken by supporters of the mystical interpretation of near-death experiences. On the one hand, there is a strong commitment to scientific research in order to understand and finally to prove NDEs. On the other hand, there is a fascination with Eastern religions and particularly with doctrines such as reincarnation. According to Morse, followers of Eastern religions such as Hinduism and Buddhism have developed a different part of the brain than adherents of Western faiths; they have developed that part of the brain that contains their "God sensor," and are ahead of Western scientists who are just beginning to catch up.

Far more interesting than the ping-pong-like debate between those who favor a physical explanation and those who favor a mystical one, though, are the theological/philosophical claims advanced by the experiencers. As in the New Age movement there is a very strong stress on self-religiosity. Near-death experiencers, on the whole, move away from identification with the religious institution of their birth. Some, it must be admitted, are led by their experience to begin attending church or synagogue, but the vast majority seem to favor private meditation rather than participation in the liturgy of the established religious communities.

The very term "experiencer" provides some clues as to why this stance is taken. Aside from charismatic church groups, the majority of Pentecostal churches and many Black Baptist churches, most denominations in North America and Europe place the emphasis on reason rather than on emotion. Emotions are welcomed as part of the worship experience, but they play a secondary role to reason. Worship is to be conducted with decorum and dignity while church business is to be managed by sound business practices. The average minister is not trained in meditation, but in systematic theology and in biblical criticism.

I can still remember with some chagrin one of the first Bible studies that I led at a church in Richmond Hill, Ontario. There was a group of ten or so adults. After we had shared come prayer concerns, I launched into the Bible study portion of the meeting. The topic, which I spent about 45 minutes exploring, was whether Sennacherib invaded Jerusalem once or twice. This was a hot topic in our Old Testament course at seminary and, as a fresh graduate, I felt that it would be a good start to our new Bible study program!

Those who have had near-death experiences find this emphasis on doctrine sterile and empty. They have caught the vision, seen the light; they are not interested in whether Sennacherib invaded Jerusalem once or 20 times. No wonder that the refrain appears that the experiencer is spiritual, but not religious. Religion is equated with reason and with dry sterile worship; spirituality is equated with a living experience of the divine.

Typical of this distinction between spirituality and religion are the comments of a middle-aged woman named Olivia who had not one, but three NDEs. Reflecting on them, she states,

I think of [the NDE] as a spiritual experience but not a religious one. I was brought up in the Jewish faith, until my father died when I was seven. After that I was brought up in my mother's faith, which was Presbyterian. So I had a very solid religious background, which I walked away from at fourteen or fifteen. I decided I was going to wave the banner and become an atheist. It didn't last very long before I started calling myself an agnostic. Then I gradually found that I, in fact, had a very strong spiritual belief – not a religious one – it's very different.[20]

Edwina, a 20-something law graduate, experienced her NDEs as a child, the first as a three-year-old who contracted typhoid fever, the second at the age of 15 while undergoing an operation for cancer of the thyroid. She is just as emphatic as Olivia is, if not more so, stating,

...I have a very strong view that church and religion are totally separated from spirituality. It doesn't matter what sort of religion. They can't help individuals along the path to understanding. The only way that people can find that understanding is by looking within themselves, and looking within and further still. Most religions I feel take people away from that. I do believe in Christ, though, and I believe in God, but not in a traditional church God. I believe that all that is light is what we call God, and Christ is a symbol of the ability of humankind to elevate themselves to that state of light.[21]

A fitting complement to this spirit of anti-institutionalism and self-religiosity is the syncretism that surfaces in many NDE accounts. For example, a woman raised in the Church of England comments, "I now feel that all religion is basically the same and I think there should be a world religion which would put an end to religious divisions and the problems that this causes."[22] Another woman named Virginia is just as explicit, noting that since her experience she now sees herself as "the younger sister" of Buddha *and* Christ.[23] One of the best-known NDErs, Betty Eadie, insists, "We have no right to criticize any church or religion in any way. They are all precious and important in his sight."[24]

Another important theological/philosophical claim advanced by the majority of those who have had a near-death experience stresses what is known as universalism. Universalism is the theory that all people will be saved and reunited with God. In the Christian tradition, this teaching is most closely linked with the writings of Origen. Origen keeps popping up throughout the pages of this book and will continue to do so. In a sense, the religious ferment evident in Western society today is an affirmation of

Origen's vision as opposed to the dominant vision of St. Augustine.

Origen was born around the year 185 in the city of Alexandria; his parents were extremely religious people who passed on their intense devotion to their son. When Origen's father, Leonides, was killed for being a professing Christian, Origen was determined to rush out in the city streets where the riots had erupted and to follow his father in Christian martyrdom. According to the story, Origen's mother who had already lost her husband and did not want to lose her son as well, prevented Origen from following through on this plan by hiding all his clothes. Apparently, Origen did not mind dying for his faith, but he didn't want to be seen naked in public and so he stayed home, no doubt in a tremendous sulk. In time, Origen became the head of what was known as the Catechetical School in Alexandria. Conflicts with the bishop of Alexandria finally led to his move to the city of Caesarea where he founded another school, which quickly became famous through the Roman Empire. In 250, during another outbreak of persecution against the Christians, Origen was imprisoned and tortured. He survived, but his health was broken and he died a few years later.

Although he was a prolific writer, many of his writings survive only in fragments or in Latin translations of the Greek originals. This is due to the condemnation which his teaching received at the hands of the established Christian church; after gaining political power, the church showed itself just as susceptible to persecuting others as once it itself had been persecuted. Origen's "sin" in the eyes of the church was his universalistic teachings. He believed in a form of reincarnation whereby human beings progressed higher or lower from one lifetime to the next. Ultimately, however, this process would come to an end in the final *Apokatastasis* when even the devil, himself, would be saved and reunited with the divine. The church condemned such teaching because it seemed to contravene the message of the New Testament; it also undercut the church's temporal authority which was based on the church's power to determine the eternal future of a person's soul. If all would be saved in the end, even Satan, then the power of the church to compel people to obey was severely curtailed.

Origen's universalism as well as that of near-death experiencers tends to support a belief in some form of reincarnation. Although experiencers may link their belief in reincarnation with Eastern religions such as Buddhism and Hinduism, the historical antecedents are found in the teachings of Origen. In Hinduism, for example, the process of reincarnation is not regarded as a series of second chances, but as an existential misfortune. Liberation is not to be found by moving through successive lives, but by escaping the cycle altogether,[25] a very different view than that put forward in the Western views of reincarnation.

Bill was a 32-year-old when he had his near-death experience as a result of his attempted suicide brought on by the misuse of alcohol. He ponders its significance some ten years later and notes,

It sure changed my ideas about death. By my religious conditioning I thought that if I didn't go to church or committed adultery, you name it, I'd burn in hell...

But now when I look forward to death, it's a pretty relaxed state. I look forward to catching up on friends. I don't think there'd be any hassles whatever, once you're over. But as far as the details go, I'm of two minds about it. I don't know whether I'll come back again after I do die, to another course of development or whatever, or whether I'll go into another sphere for an untold time (since there is no time) and when my number comes up I'll end up in Russia – I'll come back as a Russian, or whatever. I tend to believe that more than anything.[26]

One of the most intriguing assertions made by those who have had a near-death experience is the almost universal insistence that every detail in life, even the bad details, happen for a specific purpose. In the Christian tradition, this belief is known as predestination. The best known proponent of this doctrine, the Reformer John Calvin, taught that those who were saved were predestined by God to be saved. Although Calvin meant this more as a commentary on the fact that no human being can save themself and not as a philosophical commentary on life, Calvin's followers took the doctrine in directions that I doubt he would be comfortable with. The joke about the Calvinist who broke his arm and after it was over said, "Thank God that's done with," typifies the attitude of many of Calvin's followers. They taught that not only were those who were saved destined to go to heaven, but also that those who went to hell were predestined from before their birth to go to hell.

Although NDErs, by and large, do not believe in hell, but only in a hell-like purgatory, they are often every bit as fanatical about predestination as the strictest Calvinist ever was. For example, a woman named Jennifer reflecting on the psychic abilities which she feels she has as a result of a childhood NDE states,

I used to work as an usherette and I also worked in shows in town, like amateur theatricals, and I'd come and go up and down this street in Ashfield every night. All the time I was traveling at night, and I'd walk up and down the streets, and come from the trains by myself in the dark, and I was never afraid. This night I'd just gotten near my home, and there was a little lane I had to go past – I was never afraid going past this lane – and it had a door on it (someone had the funny idea to stick a door across this lane). I was never afraid, but this night I could not go past the door on that lane, I had to cross right over the road. Anyhow, the following night, an elderly lady from our boardinghouse was walking past that lane, and a fellow jumped out, bashing her up and stole her handbag. And she died from the bashing. I thoroughly believe that he was there the night I felt I had to cross the road. I've thought about why she should be killed and why I should survive, and I feel that it must have been her time to go, and it wasn't mine, and that's the only reason I was safe from it.[27]

In this connection, the life review, one of the common features of near-death experiences, forms an "aha" experience, educating the person about the events of their life and preparing them to return to life a better person. In fact, David Lorimer attempts to construct a new ethical system in his book *Whole in One* based on the life review and the lessons which it provides.[28]

A final emphasis which is common to most NDErs is the fundamental reality of love. Tom Harpur, for instance, recounts the near-death experience of a physicist whom he identifies by the initials, P.W.L. In 1965, P.W.L. had a NDE as a result of a crash on the Gardiner Expressway, a major road running along the Toronto waterfront. While he was unconscious on the operating table, where he had been taken to repair a ruptured liver as well as tears in his lower intestine, he had an out-of-body experience in which he felt that he was safe in the arms of God. "An overwhelming sense of unconditional love and concern and support completely saturated me," he later told Harpur, "in direct mind-to-mind contact, and it persisted for an indefinite duration."[29]

This emphasis on unconditional love attracts me since the New Testament makes the classic assertion that God is love, but also because the only mystical experience I had was one in which I, too, felt the overwhelming, unconditional love of God. I was on an airplane returning home from visiting a friend in Bastrop, Texas. It had been a good visit in many ways and yet, at the end, it turned bad. My last night there I struggled with thoughts of suicide which came crashing over me like the black waves off the Gulf Coast in the darkness of the early morning hours. Finally, near morning, I decided to live rather than to die. On the plane, tired and yet feeling emotional strength due to my decision not to succumb I suddenly felt the plane fade away. It was as if I was floating in space, caught between the sky and the earth, belonging to neither. Time completely disappeared. To this day, I have no recollection of whether the experience took one minute or one hour. I had no vision, but I did feel the overwhelming sense that God loved me. The only thing I remember saying to myself was, "I'm going to try to love others more." The experience must have been profound because it changed my life for the better, dramatically and instantaneously.

CONCLUSIONS

What are we to make of the near-death experiences being reported in the press, on television and radio talk shows? Many NDE researchers feel that the experiences are God's attempt to break through the scientific, secular mind set that has dominated Western culture. Thus, Cox-Chapman insists that a spiritual upheaval is at work today similar to that which occurred during Moses' day, or the birth of Jesus, or the Reformation period:

If an Ezekiel or a Muhammad were in the world now, he might make it to the newspaper tabloids. If he were media shrewd one of the talk shows might let him share his convictions. But more likely he would end up in a psychiatric unit diagnosed as delusional. So the nature of revelation had to change. It had to be diffused. Instead of one person coming down from the mountain with two tablets, we have millions and millions of people catapulted to the edge of the ineffable and back again.[30]

Margot Grey echoes this observation as she speculates that "the ever-increasing frequency of NDEs seems to be directly related to the evolutionary process of which beings are becoming ever more aware, and it could be that a higher consciousness is attempting to alert us on a collective level to the urgent need for a universal brotherhood, based on love and goodwill, manifesting in compassion."[31]

For myself, I am impressed by most of the near-death experiences I read and hear about. I find it hard to dismiss them as wish fulfillment or as a reaction to chemicals produced by the dying brain. This reductionistic approach fails to deal with the fact that such experiences cannot be willed but simply happen. Moreover, the transformation of attitudes and values which result to those who have had near-death experiences is impressive. As Dr. Bruce Greyson notes, psychiatry and therapeutic counseling take years to affect small changes in people's attitudes and behavior but "the NDE regularly brings about a total transformation almost overnight."[32] I am also impressed with the theology which surfaces in many accounts. The parallels with Origen's theological vision from people who know nothing of his writing and thought is intriguing, to say the least. Finally, the fact that various images and persons are described seems to me to confirm the validity of the near-death experience rather than, as some suggest, disprove it. If the images and person seen were always the same this would lend credence to the dying-brain explanation, the fact that they are different reflects the biblical pattern where the divine comes to us in ways we can understand.

Whether or not the increased incidence of near-death experiences (or at least the increased reporting of such phenomena) signifies that we are entering a new age is less clear to me. Interestingly, this conviction that we live in a critical time on the edge of a new age is also a feature of the New Age movement; of the Pentecostal groups, which arose within traditional Christianity; and most particularly, the latest expression of Pentecostalism – the Vineyard movement.

6

REVIVALISM

Following an information session for local clergy with John Arnott, the "senior pastor" of what was then the Toronto Airport Vineyard Church and is now the Toronto Airport Christian Fellowship,[1] a Pentecostal minister came up to me. I had directed some pointed questions to John Arnott concerning finances and the accountability structure within his church. Assuming that I was antagonistic to what was happening at the Toronto church, the Pentecostal minister, a tall, lanky man, said to me, shaking his head solemnly, "They crow like roosters and bark like dogs at that church and yet they minimize the most important gift of the Spirit, which is speaking in tongues." I nodded sympathetically and smiled to myself. I almost reminded him that according to the Bible, the most important spiritual gift was the practice of love, but he seemed so distressed that I did not have the heart.

My first exposure to the spiritual renewal taking place at the Toronto Airport Church was a rather boring introduction. I had brought a friend along who was a lapsed Anglican. He taught mathematics at a college in Kingston, Ontario, and I was interested in his opinion. Like me, he thought the experience was tame, although he took a dislike to John Arnott and felt that Arnott emanated some sort of sinister force. I myself thought that Arnott had to be the lowest key, most laid-back minister I had ever met. He looked like he was on perpetual downers. After a musician with the ever-present rock guitar led us in a few desultory choruses, Arnott and his team fielded a few questions and left. We were invited to come back that evening when the real action would start. My friend and I declined.

The Toronto Airport Church is just one of several churches which have "caught the fire." In fact, the excitement seems to have passed recently from Toronto to Pensacola, Florida, as the Brownsville Assembly of God Church in that city has been recording more visits and more responses to altar calls than its Toronto counterpart. The church sign for the Brownsville church is a religious variation of the famous McDonalds signs. Instead of billions and billions of hamburgers sold, though, the Brownsville sign reads: "Revival continues June 1, 1997, over 107,000 have responded to altar calls."

Both the Toronto Airport Church and the Brownsville Assembly of God Church belong to a rather unwieldy and diverse movement known as Pentecostalism. Pentecostalism now forms the largest Christian group within Protestantism; a remarkable feat considering the genesis of this movement lies in the first quarter of this century. It continues to be the fastest growing group as well.

ORIGINS OF PENTECOSTALISM

Two events sparked Pentecostalism. The first took place in a small Bible school in Topeka, Kansas. An itinerant faith healer and evangelist named Charles Parham, who was teaching at the school, began to emphasize the practice of speaking in tongues. Speaking in tongues had been part of the experience of the New Testament Christians, both in the form of known languages as well as in the form of ecstatic utterances that had to be interpreted to others. The apostle Paul discouraged the practice emphasizing order in worship, the priority of love, and the need for teaching which was accessible to all people (1 Corinthians 12–14). Parham disagreed with Paul's comments and felt that the practice was an important one that needed to be revived. He encouraged his students to search the New Testament for references to this practice and to be open to it. On January 1, 1901, Parham laid hands on a student named Agnes Ozman, praying that she might receive the gift of speaking in tongues. She did and the Pentecostal movement commenced.

Ozman, however, soon parted with Parham and kept her experience to herself for some five years until another outbreak of tongues took place in a small mission church in California in the section of Los Angeles now known as "little Tokyo."[2] It was initiated by a fascinating African-American preacher named William Seymour. Born in Centerville, Louisiana, the son of former slaves, Seymour moved to Indianapolis to find work and then to Cincinnati in 1900. In that city, he came under the influence of a wing of the holiness movement.

The holiness movement which sprang up following the Civil War in the United States, spread rapidly to Great Britain and Canada where it found a receptive audience among Methodists who already had responded to John Wesley's call to holy and godly living.

An important feature of the holiness movement was its insistence on the possibility of moral perfection in this life, a departure from the Lutheran tradition that the sinner and the sanctified coexist in the same person until death. Various holiness groups reacted to the Lutheran teaching, which they felt excused Christians from seeking moral and spiritual growth.

Arising out of this emphasis on the possibility of living a holy life grew another key feature which was the teaching that after baptism (the rite of the Christian's entry into the Church) there was a further step known as the second baptism. Various terms were used to describe this second step and holiness groups differed on whether it was the result of a slow process or an instantaneous change. All were agreed, however, that the initial entry of the Christian in the Christian Church wasn't enough. In part, this was brought about by the practice among many Christians of treating their faith commitment as a form of cheap grace, fire insurance for the hereafter. In opposition, proponents of the holiness movement called for a deeper level of dedication marked by surrender to work of the Holy Spirit. In effect, they separated what theologians call salvation from sanctification, claiming that the entry into the first was through water baptism and the entry into the second through spiritual rededication and surrender.

Seymour imbibed this emphasis on holiness, moving to Houston in order to work as an evangelist in association with other African-American holiness supporters. While in Houston, Seymour was exposed to Charles Parham who, in 1905, after a teaching stint at St. Louis, established a school in Houston. As he did with Ozman, Parham convinced Seymour of the importance of speaking in tongues, although Seymour himself did not receive the gift. In 1906, Seymour moved to Los Angeles at the invitation of a small holiness group in that city. His insistence that speaking in tongues always accompanied the process of sanctification resulted in his expulsion. Undeterred, Seymour set up shop in a private home where he attracted a modest following. Still, in spite of Seymour's teaching, no one had yet spoken in tongues. Then on April 9, 1906, the Spirit commenced working, the fire fell, and several people who attended the house church, including Seymour himself, began to praise God in unknown languages.

The novelty and power of this experience attracted many visitors and Seymour and his congregation moved to larger quarters, in a dilapidated building on Azusa Street. While Seymour's own leadership in the movement peaked rather quickly, countless visitors took the experience and the message of the Azusa Street Revival back home and the movement spread quickly, spawning a bewildering number of denominational groups. For the most part, these groups operated on the periphery of society for many years, working mainly with the urban poor. However, after the Second World War, Pentecostalism began to gather power and influence.

The next stage in its growth took place in 1959, when an Episcopalian priest named Dennis Bennett received the gift and began to speak in tongues. This stage has often been called the second wave of Pentecostalism. In such an analysis, the first wave began at the Azusa Street Mission giving birth to a variety of new groups called Pentecostals. The second wave was the charismatic movement, which affected mainline churches. The Vineyard organization typifies the third wave and has affected denominational groups such as the Baptists and Reformed who emphasized the primacy of correct doctrine and de-emphasized the importance of ecstatic experience.

A prolific evangelist for the gift of tongues, Bennett quickly imparted the experience to over 100 other people connected with his church in Van Nuys, California. Bennett's superiors did not like this emphasis and, in 1960, they forced him to relocate to a struggling parish in Seattle. This heavy-handed action backfired as Bennett's case received much publicity and support. Soon advocates of speaking in tongues as the mark of spiritual and moral holiness began to appear within not only Episcopalian churches but Roman Catholic churches as well.

THE IMPORTANCE OF MUSIC

One of the attractive features of both the charismatic movement and Pentecostalism is the exuberant music used in worship services. "One Sunday morning in September 1993," writes Harvey Cox, "I attended a lively Latino Pentecostal church housed in a former synagogue in what had once been the German-Jewish section of Chicago." The music was lively and the congregation of Puerto Rican worshippers were moving to the music and singing enthusiastically. As he looked around, Cox observed a small red sticker attached to a drum. From his location about three rows back he could make out the first word which was "Music" and the last word which was "Jesus." After the service, Harvey Cox slipped up to the band area to take a closer look. At last he could see the whole message: "Music Brought Me to Jesus."[3]

The birth of Pentecostalism, at least within the African-American churches, went hand in hand with the growth and popularity of jazz. In a similar fashion, both the second wave of Pentecostalism, the charismatic outbreak started by Dennis Bennett, and the third wave of Pentecostalism, were accompanied by rock and roll. Typical of the third wave is the Association of Vineyard Churches led by the late songwriter-preacher, John Wimber.

Born in 1934 in the American Midwest, Wimber had a troubled upbringing as his alcoholic father left the family when John was only a young child. In light of his childhood, it is not surprising that Wimber's own marriage experienced difficulties leading to a trial separation from his wife, Carol, in 1962. However, unlike his father, Wimber found religion and, along with his soul, his marriage was saved. Perhaps a hint of the turmoil which he went through in his childhood and early adult years echoes in his 1979 song in which he encourages his listeners to accept Christ's love: "O give Him all your tears and sadness; Give him all your years of pain, And you'll enter into life in Jesus' name."[4]

Wimber initially worked as a songwriter for The Righteous Brothers, but following his conversion began associate ministry in a Quaker church, before moving on to work at the Fuller Institute of Evangelism and Church Growth. In 1977, Wimber became pastor of a house church, which had been started by his wife, Carol, who also had experienced religious rebirth. This house church relocated to Anaheim, California, in 1982, and became part of a small group known as the Association of Vineyard Churches.

Started by a minister named Ken Gulliksen, in 1974, the Vineyard Association was composed of churches that practiced contemporary worship services for those who felt their current church services were dull and lifeless. Wimber and Gulliksen agreed on the style of worship, particularly the use of soft rock music instead of traditional hymns, but Gulliksen was opposed to the experience of speaking in tongues When he turned the leadership of the association over to John Wimber, however, speaking in tongues along with other physical manifestations of the Spirit, especially physical healings, were added to the worship experience. The Vineyard mixture proved to be a numerical success. In 1993, the Association had grown from eight to over 900 churches, and Wimber's own church numbered 5,000 members.[5]

AN OUTBREAK OF PROPHECY

A further development took place in 1988. Because of an outbreak of prophecy at the Kansas City Vineyard Church, an emphasis on prophecy was added to the previous mix of rock and roll music, healing, and tongues. The content of these new prophecies seems extremely trivial when compared with the biblical prophetic tradition, especially the eighth century B.C.E. prophets, Isaiah, Amos, and Hosea. As Jim Beverley, a long-time observer of the Toronto Airport Vineyard Church puts it,

These modern prophecies do not give God's searing rebuke to the political ideologies of our day. Moreover, they do not even speak powerfully against their own church traditions. Both in Kansas City and in Toronto, the prophets spend too much of their time defending everything at home base, while issuing the occasional spiritual death threats to critics. In short,... these modern prophets do not sound like prophets at all.[6]

Trivial as they appear to the outsider, the prophecies must sound far more profound to the insider. A prophecy given by a Canadian Vineyard pastor named Marc Dupont helped catapult the Toronto Airport Church into ecclesiastical fame. In May of 1992, Dupont had a vision in which he saw renewal breaking out in the city of Toronto, then spreading across Canada and throughout the world.

Dupont's prophecy involved flowing waters[7] as he noted, "Toronto shall be a place where much living water will be flowing, even though at the present time both the church and the city are like big rocks – cold and hard against God's love and his Spirit."[8] The revival outbreak was to touch both mainline denominational churches as well as more conservative, evangelical churches. Dupont did not specifically say that the Toronto Airport Church would be the center of this revival, although the pastor John Arnott and his wife, Carol, clearly longed for such to happen.

THE TORONTO BLESSING

Arnott attended the Ontario Bible College from 1966 to 1968, but he did not go into full-time church work. For the next decade, he worked as a businessperson. In 1979, he married a woman named Carol who had also been married previously. Together they traveled to Indonesia on a business trip and while there experienced a religious revival which convinced John to leave the world of business and begin work as a minister. Moving to Stratford, Ontario, he and his wife started a church there which became affiliated with the Vineyard Association. In 1987, Arnott started another Vineyard church in the living room of his mother's home in the city of Toronto.

Although this second church grew in size, graduating in 1990 to a strip mall near the Toronto International Airport, Arnott was still not satisfied. He and Carol wanted something more – they wanted something big and something grand to happen. After a fall healing meeting led by Arnott's friend Benny Hinn, John and Carol devoted mornings to praying for revival to hit their church. Nothing happened but they did not give up. When they heard of a revival going on in Argentina associated with a Pentecostal minister named Claudio Freidzon, they determined to visit Argentina and to catch the Spirit. Freidzon prayed for John and John fell to the ground. Later Freidzon asked him whether he wanted the anointing of the Spirit, so he could have this impact on other people. Arnott replied that he did. Freidzon "slapped [his] outstretched hands,"[9] and like a form of Roman Catholic priestly succession, Arnott received the power. However, when he returned to Toronto still nothing spectacular happened in his church so Arnott began to invite speakers whose ministries seemed to be experiencing spiritual manifestations.

One such church was the St. Louis Vineyard Church pastored by Randy Clark. Arnott invited Clark to hold a series of revival meetings at his church. On January 20, 1994 the Spirit finally struck. Arnott was so excited that he refused to let Clark go home and instead paid for Clark's wife to fly up to Toronto, so that the meetings could continue unabated. Strange physical manifestations began to occur. People began to burst into gales of laughter, they fell on the floor jerking (known as doing carpet time), they staggered around as if drunk, they began to experience visions, and most popular of all, they began to experience physical healing.

By 1995, the revivals had become so well known that people were flying to Toronto from various countries in order to participate in the services which were held six nights a week. These visitors took the message and the experience back to their home churches and the revival began to spread. The secular media, first in Great Britain and later in Canada, became interested in what was going on. One observer, Robert Hough, describes a visit to the Toronto church during a "Catch the Fire" conference. The conference was scheduled to last for three days. Hough thought that he would have to stay the entire time waiting for the climax, for the manifestations to start. The conference had hardly gotten underway, however, when the Spirit struck:

Then, it happened. The man sitting beside me, Dwayne from California roared like a wounded lion. The woman beside Dwayne started jerking so badly her hands struck her face. People fell like dominos, collapsing chairs as they plunged to the carpeting. They howled like wolves, brayed like donkeys and – in the case of a young man standing near the sound board – started clucking like a feral

chicken. And the tears! Never have I seen people weep so hysterically, as though
every hurt they'd ever encountered had risen to the surface and popped like an
overheated tar bubble. This was eerie, badass stuff – people were screaming,
their bodies jerking unnaturally, their faces contorted with tics. Yet the most
unsettling were the laughers, those helpless devils who were now rolling around
on the floor, holding their stomachs, their minds gripped by some transforma-
tive, incomprehensible power.[10]

In time, the events such as Hough describes, as well as the prominence
given to them, resulted in a break between the Association of Vineyard
Churches and the Toronto Airport Vineyard Church. In particular, it seems
that the animal noises were a focus of much concern by John Wimber and
the Vineyard leaders. In fact, even Rodney Howard-Browne, a South
African-born evangelist and healer who figures prominently in the revival
movement associated with Toronto and elsewhere, stated in an interview,
"We don't have any barking or roaring in our meetings. If you bark like a
dog, we'll give you dog food. If you roar like a lion, we'll put you in the zoo.
If you cluck like a chicken, we'll give you bird seed."[11] In spite of initial
harsh feelings, antagonism on both sides was soon patched over and the
Toronto church continued on, pausing only to change its name.

In fact, the Toronto Airport Christian Fellowship as it became known
after Wimber withdrew his support, is supposedly on the verge of a new
outbreak of the Spirit which will make the previous manifestations seem
tame. At the third anniversary meeting on January 20, 1997, Randy Clark
was invited back as guest speaker. No sooner had he announced the title of
his message, "The Making of a Prayer Warrior," than Carol Arnott, now
listed along with her husband John as Senior Pastor of the church, fell to
the ground and began slashing at the air holding her hands together. For
about 20 minutes Carol did this, until Randy finished his message, at which
point she stood up and delivered the following message from the divine:
This is My sword, this is not man's sword, this is My golden sword. The ways you
have been using My weapons, the methods you have been using in the past, you
are to throw them away because I am giving you My sword now and the old ways

of doing things will not do. The old methods will not be acceptable to Me anymore because I am doing a new thing. Do not look to the yesterdays but look to the future because I am doing a new thing and this new way is not the old. This new way is new and you must throw away the old ways of doing things and take up My sword because My sword is made of pure gold and is purer and is mighty. If you wield it the captives will be set free, the chains will be broken and the healings will be manifest because it will not be by might, nor power, but by My wonderful Holy Spirit. It is by Him, it is by Him that this new wave will be brought forth, it is by Him that the King of Kings and Lord of Lords will rise again.[12]

Whether Carol's prophecy will see the limelight return to Toronto or not is still unclear. Revival meetings continue unabated, but the main activity of the Spirit seems to have moved to Pensacola, Florida. The catalyst for this outbreak of revival was another prophecy, this time delivered by David Yongii Cho, the minister of a church in Seoul, Korea, which claims a membership of some 800,000 members. In 1993, David Cho, apparently unaware that the Spirit was to fall on the city of Toronto, lamented the religious situation in America at that time. He prayed and asked God if revival was going to by-pass America. The answer he received was to go to a map and put his finger on it. He did this and looking down noted that his finger was placed on the city of Pensacola.

Not having been to the Pensacola church it is difficult for me to compare its operation with the church in Toronto. From written reports on its Web site, it seems that the Pensacola revival is somewhat more exuberant than its Canadian counterpart. Apparently, the Holy Spirit is governed by cultural differences, which allow Americans to be more boisterous than their Canadian cousins!

THE NEW AGE MOVEMENT IN DISGUISE?

In his excellent book on Pentecostalism, Harvard scholar and religious observer Harvey Cox claims that as scientific modernity and conventional religion lose their "ability to provide a source of spiritual meaning, two new contenders are stepping forward – 'fundamentalism' and, for lack of a more precise word, 'experientialism'."[13] In the 1980s, Cox made a similar observation about the future of fundamentalism and liberation theology.[14] Clearly, he has a penchant for setting up contenders for the religious future. Nonetheless, I believe that he is onto something. Rather than the simple contrast of fundamentalism and either liberation theology or experiential Pentecostalism though, I believe that more exotic brews of religious movements are arising in reaction to the sterility of the Western way of life. These, as I've already implied, include New Agers with their crystals, modern-day shamans practicing a variety of alternative medical treatments, near-death experiencers, Wiccan feminists, Gaian worshippers, and Pentecostalism, especially in its latest revivalist clothes.

It should occasion no surprise that one of the constant criticisms directed against the Vineyard movement, the Toronto Airport Church and other revivalist preachers and groups is that they are New Agers in disguise. Within conservative Christian circles, this has been the kiss of death. Everyone there knows that the New Age is a satanic movement sent by the archdevil himself to enslave the souls of many. While not supporting such a view, I admit that there are valid points in the criticisms made by conservative Christians against the events taking place at Toronto and elsewhere. Yet, rather than viewing what is taking place in negative terms, I prefer to adopt a positive stance. I am not convinced that the religious ferment evident in many different out-workings is the final answer to the spiritual emptiness of post-World War I society, but it is a necessary counterbalance.

Certainly, the similarities between the movements described earlier in this book and the Vineyard-Revivalist movements are numerous and clear. The practitioner of therapeutic touch may use a form of pop-Hinduism to justify their practice, while the pastors at the Toronto church use biblical data, but the change in perspective from depicting the human being as a biochemical machine to a vibrating energy field is common to both. Indeed, even the stress on the new physics, which fuels much of the practice of alternative medicine as well as the New Age movement, is present within conservative revivalism. As one fan of the revivalist movement, William DeArteaga, asserts,

One can note the vast difference between our hermeneutic (Christian faith-idealism) which is buttressed with the analogies of quantum physics and the hermeneutics of Christian materialism based on the science of the eighteenth century. Whereas the understanding of material science forced a radical discontinuity between spiritual activities and the material order, those discontinuities have now been evaporated. The spiritual order can now be understood to operate in harmony, not in contradiction, with the fundamental laws of the universe. The analogies between quantum physics and the spiritual life reveal a continuity of intention in the mind of the creator. God intended the universe to be spiritual – from subatomic particles to archangels.[15]

While I wonder what Werner Heisenberg would think of such a statement, not to mention Stephen Hawking who has no tolerance for religious interpretations, clearly the new physics generates excitement in both the New Age and revivalist communities. The list of essential beliefs within this new wave of Pentecostalism sounds suspiciously similar to observations made earlier in the book concerning New Age and near-death teachings, but there are some interesting differences.

DIFFERENT VIEWS ON THE MILLENNIUM

For example, a key similarity is the emphasis on the critical historical moment in which we now live. The difference lies in the pessimistic view exemplified by Pentecostalism and the more optimistic view held by the various New Age groups. Pentecostals, along with conservative Christians, feel that the end of the world will be inaugurated by a time of great tribulation as outlined in Revelation, the last book of the Christian Bible; New Agers look forward to a golden future here on earth, to an Age of Aquarius. Even this is changing, however, as New Agers, especially those who have experienced NDEs, are reporting more somber visions of the future while the new Pentecostals seem to be shifting from a premillennialist view to a more optimistic postmillennial view. In fact, Hank Hanegraaff, a conservative Christian writer, criticizes the Toronto Airport Church pastors for predicting a great revival as the precursor of the return of Christ and the end of the world. According to Hanegraaff, "in biblical eschatology, the precursor to the coming of the Lord is great apostasy" not ecstatic revival.[16]

This millenialist strain is a common feature in the various movements analyzed in this book. In light of the psychological pressure created by the end of the second millennium and the dawn of the third, this comes as no surprise. It is also not new. Christianity has always had a millenialist tinge that surfaced in various groups. As mentioned, New Age teaching parallels Joachim De Fiore's division of history into three ages – the age of the Father, the age of the Son, and the age of the Spirit. Harvey Cox claims that Pentecostalism also reflects this division. Commenting on an 1894 claim by an African-American writer, James Theodore Holly, in which he postulated the existence of three ages, Cox notes:

When I read this bold prophecy made by a black American Christian over a century ago, I was reminded of another seer. In the closing years of the twelfth century an Italian Cistercian abbot named Joachim of Fiore propounded a theology of history that bears certain intriguing similarities to Holly's, though Holly never mentions Joachim...

In this dawning new age, Joachim taught, the church would no longer need a hierarchy because the luminous presence of God the Spirit would suffuse all peoples and all creation. Priests and bishops – even popes – would become superfluous. Sacred scriptures would no longer be necessary because the Spirit would speak directly to each person's heart. Further, with the arrival of this new age of the Spirit, the strife and hostility that had divided Christians from infidels would disappear. All clans and nations would be joined in a single harmonious body. It almost seems as though Joachim was writing the script for the pentecostal drama that would come 700 years later.[17]

Much earlier than Joachim, however, the prophet Jeremiah looked forward to a similar development as he predicted that the day would soon dawn when God would enter into a new relationship with the people of Israel. Instead of the law being written on stone tablets and taught by a special class of religious leaders, God's law would be placed within the individual, written on the tissue of the human heart. In Jeremiah's words, "I will put my law within them, and I will write it on their hearts; and I will be their God, and they shall be my people. No longer shall they teach one another, or say to each other, 'Know the Lord,' for they shall all know me, from the least of them to the greatest, says the Lord" (Jeremiah 31:33b–34).

INDIVIDUAL RELIGIOSITY

Accompanying the omnipresent millennial vision, whether interpreted optimistically or pessimistically, is a stress on individual religiosity and a strong anti-establishmentarian mentality. This is muted in Pentecostalism, since by definition churches are community organizations often led by strong, charismatic leaders. However, the rhetoric that surfaced in our discussion of NDEs persists even within Pentecostalism. Isaac, a guitarist in the Brownsville Assembly of God youth band, gives a typical exhortation. At a rally held in Baker, Florida, he stated that "young people and adults, should be united... there were no doctrines going to Heaven... [People] needed to drop all religion and join together to lead dying souls to Christ."[18]

More pointed is a comment from Julie Suggs as she reports on a revival in Valdosta, Georgia, initiated through a preaching series by the outreach minister of the Brownsville church. Scheduled to be a three-day meeting, the revival had moved into its sixth week when Suggs reported, "To date almost 80 people have been saved, many healed, delivered, set free from addictions, set free from bondage, *set free from religion*." [emphasis mine][19]

Even the theme of self-divinization is present. Typically, the power within the individual is attributed to the Holy Spirit, dwelling in the believer. However, Kenneth Copeland, an evangelist who had strong connections with Rodney Howard-Browne, one of the key players in the revivalist movement, delivered the following prophecy from Jesus:

Don't be disturbed when people accuse you of thinking you're God. Don't be disturbed when people accuse you of a fanatical way of life. Don't be disturbed when people put you down and speak harsh and roughly of you. They spoke that way of Me, should they not speak that way of you?

The more you get to be like Me, the more they're going to think that way of you. They crucified Me for claiming that I was God but I didn't claim I was God; I just claimed I walked with Him and that He was in Me. Hallelujah. That's what you're doing.[20]

The final element in this emphasis on self-religiosity is the elevation of experience over doctrine. This is evident in its starkest form among NDErs, but also very noticeable within the Pentecostal tradition and its most interesting current expression – the revivalist movement. Indeed, Harvey Cox uses the word experientialism as a synonym for Pentecostalism. As with NDErs, the experience of the infilling of the Spirit is so overpowering that rational analysis seems a trivial and unimportant impediment, rather than an aid to true spirituality.

CLERGY "REVIVAL"

This overwhelming power of the experiential may well account for the high number of clergy who visit the revivalist churches in Toronto, Pensacola and elsewhere. At first, it seems surprising that so many ministers make the pilgrimage and have the experience. Society at large views ministers as people who are already spiritual and in touch with the divine. This is not a true picture for a couple of reasons. The first is that turmoil in the Christian church in North America has resulted in enormous psychological pressure and stress being placed on the typical minister. Burnout, sexual misconduct, dropout, and alcoholism and other forms of drug abuse are much more prevalent than many religious officials care to admit.

Typical of such people is a friend who visited me. We had not seen each other since theological school and so it was catch-up time. Sitting on a green chair on my side lawn, he updated me on his life story. His marriage had ended; his wife was now happily remarried, living in Texas with their two children. He had continued in ministry, but even the confines of the United Church were too claustrophobic for him. The worship of the church made little sense, the liturgy seemed archaic. He complained that the Baptist emphasis on separation of church and state that dominated in Canada and the United States resulted in the trivialization of the role of the minister and the place of the church. Finally, he left pastoral ministry and was now studying to be a doctor. As I studied his face, aged with tension, and noted his long ponytail, I couldn't help wondering what would have happened to him if instead of leaving the Pentecostal church to study at a Baptist seminary, and then moving on to a United church, he had stayed a Pentecostal. In spite of his keen mind and philosophical bent, would he have ended up on the carpet, struck down by the Spirit?

That is exactly what happened to another ministerial acquaintance of mine, Doug Coombs. Doug's story appears in the introduction to John Wimber's book on healing and so I feel free to use his real name. After ministering in Kingsway Baptist Church in Toronto for 14 years, Doug was worn out and spiritually empty. A visit to his brother Wayne in California seemed like much needed therapy. It proved to be that and more as Doug's brother had signed him up with his wife Mary to attend a "Signs and Wonders Church Growth Conference" led by John Wimber. At first, Doug and Mary were appalled by what they saw going on, but then on the Thursday night of the conference, Doug had the experience. At the end of the meeting that evening, John Wimber called the pastors forward for prayer. Out of 3100 attendees, about 1000 were ministers. Drawn forward, almost against his will, Doug heard John Wimber pray for him and the other ministers. Then it hit. He recalls:

I was knocked over into the arms of a huge man who, I later learned, was a professional football player with the New York Giants. He said, "You are a pastor from Canada who has just resigned from your church. The Lord has called you to a new church, he will add many years to your life, and he will give you the gift of evangelism." There was no way he could have known that I was a Canadian or that I had just resigned from my church. While he was speaking these things I felt a warmth throughout my body and for the first time I experienced the joy and peace of God. I was delivered that night of the anger, cynicism, and bitterness that I had allowed to take root deeply in my heart, and that was holding me back in my walk with God.[21]

Whatever is happening to such ministers, it feels good. Set free from cynicism, they once again vibrate to the touch of the Spirit, deep within.

Along with ministerial burnout, another important factor in the prevalence of clergy going forward at revivalist meetings is the emphasis of ministerial training. Enchanted with the university model, seminaries seek to impart book rather than heart knowledge. They assume the heart knowledge is already present and that what is needed is book knowledge. This means that many ministers have never felt any connection with the divine, although they can cite doctrine *ad infinitum*. Fortunately, this is changing. My own training included retreats plus a year-long course in clinical pastoral education. Designed to help me function in my counseling responsibilities, CPE, as it is affectionately known, also served as a challenge to my spiritual life, helping me to look within myself and thereby find the divine which lies in all of us, if only we are open. I still function more out of my head than my heart, but that seems to be my style and personality as a quiet person who loves nothing more than to read a good book. For many ministers, though, seminary distorted their personalities, elevating reason over emotion. Little wonder that so many go forward, fall on the carpet, and find their heart's joy once again.

In this regard, even if it is simply coincidental, it appears to be significant that after the break with the Association of Vineyard Churches, the Toronto church dispensed with the large, fundamentalist-oriented statement of faith shared with other Vineyard churches. In spite of being attacked for elevating experience over sound doctrine, the Toronto church adopted a very simple statement of faith, dispensing with statements such as "we believe... that the Bible is without error in the original manuscripts... our final, absolute authority, the only infallible rule of faith and practice." Instead, one finds a simple sentence: "We believe that the Bible is God's Word to the world, speaking to us with authority and without error."[22]

HEALING

The most overt connection with what is happening in the New Age movement, with NDErs, and with practitioners of alternative medicine, concerns the emphasis on healing. Randy Clarke's revival meetings are credited
with sparking the Toronto revival, but what really galvanized public interest
was an account of the healing of a young girl named Sarah Lilliman who
was in hospital in a near-blind and vegetative state. Prompted by a vision
that she experienced at the church, Sarah's friend went to visit her in hospital. According to Vineyard accounts, Sarah's sight returned and in a few
short hours, she emerged from her coma. Although subsequent study has
shown this account to be stretching the truth (to put it kindly) the initial
report of the healing caused an enormous sense of excitement and energy.[23]

Margaret Poloma, an American sociologist who has written previously
about the Pentecostal movement in the United States, published a short
document on the Internet entitled "By Their Fruits: a Sociological Analysis
of the 'Toronto Blessing'." While it is abundantly clear that Poloma is not
an unbiased observer, her statistics are, nonetheless, worth noting.

*A holistic process pointing to the interdependence of spirit, mind and body appears
time and time again in the testimonies of visitors to the Toronto Airport Church.
These testimonies provide some flesh for the skeletal statistics which show 78 percent
of respondents had experienced "an inner or emotional healing," 5 percent had "experienced a healing from clinically diagnosed mental health problems," and 21 percent reported "physical healing" as a result of the Toronto Blessing.[24]*

In a *Christianity Today* sponsored symposium, John Wimber explains that
healing seems so unusual and exotic because "we're twentieth-century ma-
terialists, we're rationalists, so we look for natural, materialistic clarifica-
tions of things."[25] He then goes on to relate a story of a meeting in Melbourne,
Australia, where the Lord spoke to him about the health problems of a
woman in the audience. Wimber then spoke up and told the gathering that
there was a woman with a cleft palate who had had two operations to try to
fix it, during which her teeth were removed. After diagnosing her problem,

Wimber added that his knowledge of the woman's condition was a sign that God would heal her palate. In reaction, a doctor in the audience became concerned and admonished Wimber. Put simply, the doctor was on the side of conventional medicine while John Wimber was engaging in alternative medicine.

Even the methods used, apart from prayer, sound similar to practices that occur in alternative medicine. Therapeutic touch is really a misnomer since the practitioner does not touch the person but manipulates the energy field, which supposedly radiates from the body. Whether or not he is aware of therapeutic touch, a revivalist named Chuck Schmitt should set out his shingle and set up shop. He notes, "When I pray for people, I have them show me their right hand. I place my hand above theirs, not touching them. I ask God to show them His power in confirmation of the Gospel message I just preached to them. People feel power, like static electricity, or wind, or heat, or a magnetic force. They feel tingling."[26]

In connection with this emphasis on healing, mention should be made of the "manifestations" occurring at the revivalist churches. Using similar terms as Chuck Schmitt, Arnott explains that people shake and fall down at his meetings because they have been struck by the power of God's Spirit. He asks, "Suppose I handed you two bare, electric wires and said, 'Would you please hold these tightly? Taken one in each hand and hold while I plug it in.' How many of you think you might shake a little bit?"[27]

As people are prayed for at meetings, they exhibit strange physical reactions. It is as if the divine were performing a celestial version of shock treatment on them. Following the experience, a significant number relate that they feel emotionally better and mentally more alert, in spite of what to an outside observer appears to be a fit of madness. Another ministerial acquaintance who has also written about his experiences at the Toronto church, Laurie Barber, describes his experience this way:

As one service progressed, I began to have heaviness in my body, particularly in my hands and forearms, like one feels after trimming the garden hedge for an hour or so with old non-electric hedge trimmers with the resulting sense of heaviness and trembling, as slightly used muscles slightly spasm. More recently, a vibrating in my lower jaw and the stem of my tongue occurs when I am being prayed with or when I am praying for someone else.[28]

In spite of these rather unsettling physical feelings, the result was an emotional healing and sense of empowerment in Laurie Barber's life and ministry.

OTHER MANIFESTATIONS

However, not all the manifestations can be explained as psychological or physical healing. The spectacle of people getting on their hands and knees and barking like dogs, one fellow even going so far as to lift his leg to urinate, as well as people roaring and crowing has nothing to do with healing. These activities, which must provide some form of therapeutic release, are treated by the staff at churches such as Toronto as prophetic signs. The lion noises are meant to signify that the lion of Judah, a reference that Christians interpret messianically as applying to Jesus, is coming in power and glory. The crowing noises supposedly are a sign that revival is here and the end of the world is at hand.[29]

Ecclesiastically speaking (if you can forgive me), what I find interesting about the various physical manifestations – drunkenness, shaking, jerking, falling down, trembling, laughing, crying, animal noises and the like – is that they function as a sign that God exists and is involved with humanity. To use theological terminology, they form a sacrament of manifestations, which to a nonverbal, television-oriented society such as ours, may be the most effective way of communicating the presence of the divine. The original sacrament of God's presence with God's people was and is the Mass. In the bread, which on the lips of the receiver becomes the body of Christ, and in the cup, which is the blood of Christ, the believer knows that God is present. That is why in a Roman Catholic church where this view is held most deeply, the priest will bow when he passes by the altar.

The Protestant Reformation exchanged the sacrament of preaching for the sacrament of the Mass. The outward and visible sign of the presence of the divine was no longer the bread and the cup, interpreted in many churches such as my own as mere symbols, but the spoken word. Even a bad sermon – and many were and are – signaled that God was present, communicating and caring with humanity.

As the television image has replaced the written and spoken word, however, a more visible and dramatic sign is needed. Arnott puts this in crystal-clear terms, even though he would not analyze things as I have, when he praises his wife for her spiritual openness by stating, "if you were having a rather bad day and wondered if God were still with you, you could pray for Carol and she would receive wonderfully and probably fall down. *You would feel better, knowing that God was still with you and answering prayer.*"[30] Whether God caused Carol to fall down is unimportant. The most important thing is that one believes the physical manifestations form a dramatic sign of God's presence; they become sacraments.

THE INTERNET

One other feature of the revivalist movement within Pentecostalism should be noted. As evidenced by the number of citations from Internet documents in this chapter, this revival is Internet driven. How the Internet will shape the revival I do not yet know, but having been a recipient of countless e-mail letters from Richard Riss as part of the Awakening List, I can personally testify to the impact of the Net. Stories that would circulate far more slowly and perhaps peter out in the past are now flashed instantaneously around the world. In fact, on a site connected with the revival in Pensacola, you can click on an image map of the United States to be connected with churches that are involved in revival. In a quintessential caricature of American hubris, each of the states is shown while the revivalist churches outside the United States are linked under a small icon entitled "the rest of the world."

CONCLUSIONS

What is one to make of the events happening at Pensacola, Toronto and elsewhere? There are at least four options. The first is that the people experiencing these manifestations, healings, visions, and prophecies are fakes. While I have no doubt that some are, it would be stretching the imagination to claim that thousands upon thousands are faking experiences for personal gain.

The second is that a profound sense of social and personal dislocation predisposes some people to a form of hysteria or hypnosis. After reading testimony upon testimony this explanation rings true. The high percentage of ministers attending these revival meetings gives an important clue. However, this explanation begs another question. Why are so many people experiencing social dislocation? The answer to this question points to the large social and ideological changes taking place. We are, it seems to me, at the end of one era groping to find our way, safely and happily, into the next. Whether this should be described as a "new age" is often a question of semantics, a contrast between those who favor continuities over those who favor the novel and the new.

The third is that people such as Rodney Howard-Browne, John Wimber and others have stumbled on to some sort of power which does manipulate a person's energy field, either to produce healing or to produce manifestations. This presupposes that people do have an energy field and that the Spirit functions on a level not yet known to science. Such power could be used for good; it could also be manipulated for selfish and destructive means. After reviewing the material in support of alternative medicine, as well as my own experiences, I believe that there is some form of energy transfer which has not yet been discovered by science, and which may never be.

The fourth and final explanation is that these revivalist outbreaks are the work of the divine. Such an explanation presupposes that the divine exists and seeks to relate to us, something that many practitioners of alternative medicine, scientists, and Gaian supporters would deny. Clearly, I take my stand on the side of the existence of a personal, loving, involved God and so am predisposed to agree with this explanation. However, Luther's theology of the cross, whereby the divine is hidden in suffering and weakness, as well as that haunting passage from the Hebrew Scriptures where Elijah finds the voice of God not in the fire, or the wind, but the still small voice within,[31] prompts me to be skeptical of much that goes on in revivalist churches. There is a will to believe which overrides questions of a more rational nature. People cannot stand the vacuum of meaning that we live in and so they rush to churches like the Toronto church without adequately pausing to think things through.

In the end, perhaps all four explanations are right. In New Age and Pentecostal circles there is fakery for personal gain, whether that gain be financial or in terms of prestige within the community. But there is more to it than that. The times they are a-changing. I hope God is behind these changes. I hope they will result in a more loving, less materialistic society conscious of the activity of the divine, open to the mystery of God, and concerned about the life of the Earth. As one whose passion has been social justice issues, though, I also hope that these changes will not just touch individuals in the West, allowing us to live at ease in Zion, but will reach out to the world to enshrine peace and justice more fully. In their own ways that is what the next three movements, fundamentalism, liberation theology, and feminism seek to do.

PART THREE

THE THEOLOGICAL TERRAIN

7

FUNDAMENTALISM

Stephen Cohen[1] did not look Jewish but Russian, not that this would be out of place since Stephen's parents were born in Russia before immigrating to Israel in the late 1970s. At the age of 16 Stephen was uprooted from his family and friends and moved to the city of Tel Aviv on the coast of the Mediterranean Ocean. Later, Stephen attended university in the city of Haifa a bit further north. It was there that he came in contact with members of the radical settler group known as the Gush Emunim (Bloc of the Faithful) led by the Jewish activist Daniela Weiss.

Entranced by Daniela's concept of "Messiah-time" when Israel would be restored to its former greatness both politically and religiously, Stephen moved to the West Bank. There his parents lost contact with him until January 1984, when they were called to the morgue at the Jerusalem hospital to identify the remains of a young man who had been tentatively identified as Stephen Cohen. As the pathologist lifted back the thin sheet to reveal what was left of Stephen's body, his father turned and threw up in his hands, while his mother stood dead still, in shock.

Except for his face which, surprisingly, had not been badly scarred by the bomb blast, there was not much left of Stephen. He had been too close to a bomb which he and other members of his group had been building to plant under the Dome of the Rock in the city of Jerusalem. They were to begin work on a tunnel and were transporting the bomb from a safe house in the West Bank to Jerusalem when it blew up, killing Stephen, and two others.

They never got near their objective of destroying the Dome of the Rock, the third holiest Muslim shrine.

Situated in the middle of the old city of Jerusalem, the Dome of the Rock is a beautiful building. Inside, on the roof of the dome, are writings from the Koran in ornate calligraphy since the Muslim faith, like the Jewish, prohibits the use of images within religious buildings. Underneath the dome is a large flat rock, by tradition, the location where Abraham lay his bound son, Isaac, in order to sacrifice him to God. On the outside, gleaming gold caps the top of the dome.

The problem with the Dome of the Rock, at least for people such as Stephen, is that it sits right in the middle of the Temple Mount. By its presence, it prevents the rebuilding of a new Jewish Temple. It was this, rather than the fact that it is a Muslim shrine, which prompted Stephen to attempt what he did. The Messiah could not come back and Israel could not become the pure religious state it was meant to be as long as the Temple remained in ruins. The Temple could not be rebuilt until the Dome of the Rock was removed. In the end, the decision was easy.

Secular Jews, Stephen's parents found it hard to understand how their son could turn to a form of religious faith which has been called fundamentalism. In this, they resemble many other parents whose children have rejected the liberalism of their parents and embraced a fundamentalist form of their respective faith.

Fundamentalism, especially in its North American dress, is a topic with which I am very familiar, as I did my doctoral dissertation at McGill University on one of the key fundamentalist leaders in North America during the first half of this century. The name of Thomas Todhunter Shields is not well known these days but once he was a force to be reckoned with not only in his home country of Canada but also in the United States, Great Britain, and Australia. A controversial figure, he was the flashpoint for a bitter denominational split within the Baptist movement in Canada. As a consequence, his papers were kept sealed in two vaults in his home church, Jarvis Street Baptist Church, in Toronto. Researchers who asked to see his sermons and letters were turned away. Whether I was just lucky or whether I managed to work a minor miracle as the late Queen's history professor George Rawlyk has suggested,[2] I was able to access the Shields' material at the Jarvis Street Church.

For a historian it was a treasure trove. Many of the sermons were beautifully crafted, as Shields was known as a pulpiteer. It was the letters, however, which proved to be the most interesting. In correspondence with almost every prominent fundamentalist leader in North America, as well as important political figures in Canada, the letters provided a fascinating insight both into Shields himself, and into the fundamentalist movement which he helped form. They gave flesh to the sermons.

On one occasion as I picked up a letter, bits of red plastic fell out onto the table. Thinking that the letter had been sealed with wax and that the wax had now disintegrated my pulse began to beat faster as I reasoned that I was going to see an important document. When I gently opened the letter, the red plastic bits proved to be the remnants of a 40-year-old condom. Inside was an unsigned hate letter telling Shields that rumor had it that he had impregnated so-and-so and in the future would he be so kind as to use one of these.

Another letter that piqued my interest was one that Shields wrote but seems never to have sent. Shields had been drafted as the chancellor of what was to be the flagship of the fundamentalist movement in the United States. A small university in Des Moines, Iowa had gone bankrupt and several fundamentalist leaders reasoned that if they could build up a strong fundamentalist university they could compete with the liberals on their own terms. Shields was asked to be chancellor and agreed. Because of his busy schedule he sent his secretary, a single woman named Edith Rebman, to oversee the work. She soon came in conflict with the principal of the school, H. C. Wayman, who resented having to report to a woman. Tensions between the two boiled over and charges were brought against the principal that he had bogus academic credentials. Shields traveled to Des Moines to investigate.

He was not greeted warmly when he arrived there, as the majority of students seemed to sympathize with their disgraced principal. Long before

Kent State and the Vietnam protesters made campus riots front-page news, Des Moines students led the way. At an assembly, Shields made a group of American fundamentalist students sing *God Save the Queen*; this proved to be too much and they tore the school apart forcing Shields and his secretary to hide in a washroom until the police finally arrived to escort them to safety. The rumor that Shields and his secretary were having an affair was spread far and wide.

Whether or not this accusation was true, Shields was extremely sensitive to its stain on his reputation and possible impact on his employment as a minister. When a friend of his, William B. Riley, the fundamentalist pastor of the First Baptist Church of Minneapolis, raised a criticism concerning Shields' involvement with his secretary, Shields saw red. He drafted a blistering letter in which he threatened to name the names of several women associated with Riley, if Riley did not back off. The letter was never sent, and Riley and Shields later patched up their differences.[3]

NORTH AMERICAN FUNDAMENTALISM

Fundamentalism *proper* began in the United States and Canada immediately following the First World War. The word comes from a 1920 article written by Curtis Lee Law, the editor of a religious journal called the *Watchman-Examiner*. Some five to ten years earlier, immediately before the Great War, a series of 12 pamphlets were produced and funded by Lyman and Milton Stewart. Using money gained from the sale of oil, the booklets were distributed free of charge to every English-speaking pastor, missionary and theological professor in the world. They served as an attempt by the Stewart brothers and others to articulate and defend the central core of Protestant doctrine. Each booklet carried the same title, *The Fundamentals: A Testimony*. The writers and the positions taken, however, were not yet fundamentalistic, in that while exposing conservative theological doctrine they did not call for ecclesiastical separation or for militant action against fundamentalist opponents. Instead, they were the warning shots by conservative supporters concerned about the direction of North American Protestantism.

The outbreak of the First World War put issues of theological controversy on the back burner. Conservative church leaders were accused by Chicago professor Shirley Jackson Case of being in the pay of German interests since their premillennial doctrine undercut human efforts to better society.[4] Determined to show that they were not disloyal to their country, such leaders joined in enthusiastic support of the war effort. In time, they came to see the war as a great catharsis which would cleanse American society of its theological liberalism and would result in the growth of the Christian church.

This return to conservative Protestantism failed to happen. Instead, the First World War accelerated the secularization and urbanization that threatened Protestant hegemony in North America. Margaret Hand spoke for many when she stated, "It was the First War that shot religion high, wide, and handsome, as far as a great many of us were concerned. We just lost faith in it. I mean, after all, you could pray your head off but it didn't mean you were going to have your men come home to you."[5] Rather than dealing with the challenge of how a good and loving God could allow atrocities such as war and tragic death to happen, fundamentalist leaders laid the blame for the diminishing role of conservative Christian faith in North American society at the feet of modernistic teaching. They called for a theological war to accomplish what the First World War failed to do.

For a while, the fundamentalists appeared to be winning. Their more liberal opponents complained that the label "fundamentalist" was unfair. It gave undue advantage to the fundamentalists since every Christian believed in upholding the fundamentals of the faith. Some religious careers were ruined and the early 1920s looked very promising for the fundamentalists. Three events changed public opinion, however, and along with other important factors, pushed the tide of the theological battle in the direction of the liberals, or modernists, as they were then known.

The first and the most important event was a court trial which took place in 1925 in Dayton, Tennessee, known as the "Monkey Trial." A young high school science teacher named John Scopes began to teach that Darwin's theory of evolution was the correct scientific view and that, consequently, the biblical view of a seven-day creation depicted in the Book of Genesis was scientifically invalid. Because this was contrary to the state law at the time, Scopes was brought to trial. His plight became a national issue when his case was taken up by a New York lawyer, Clarence Darrow. To fight fire with fire, the prosecution of the case was given to a former candidate for the presidency of the United States and a committed conservative Christian, William Jennings Bryan. The contrast could not have been more striking.

In popular imagination, there were on the one side the small town, the backwoods, half-educated yokels, obscurantism, crackpot hawkers of religion, fundamentalism, the south, and the personification of the agrarian myth himself, William Jennings Bryan. Opposed to these were the city, the clique of New York-Chicago lawyers, intellectuals, journalists, wits, sophisticates, modernists, and the urbane agnostic Clarence Darrow.[6]

In the end, William Jennings Bryan won the court case, but due to the writing of influential journalists, such as H. L. Mencken, the victory came at the cost of heavy losses. William Jennings Bryan was made to look like a fool and fundamentalism was discredited.

The following year, another event harmed the fortunes of the fundamentalists. A Baptist preacher in Fort Worth, Texas, J. Frank Norris, launched a series of sermons attacking Roman Catholics. The growth of the Roman Catholic church was seen by many fundamentalists as a root cause for a slippage in the religious tenor of the United States. Since the mayor of the city was Roman Catholic, Norris' sermons had political implications. One day, the aide to the mayor came to Norris' office to plead with him to stop his attacks. The story diverges from here on. According to Norris, when he refused to back off from his anti-Catholic attacks, the aide to the mayor walked toward the door. Instead of leaving, though, the aide turned around in a threatening fashion. In Texas, this could mean only one thing to Norris, and so rather than be shot, Norris pulled a gun from his desk drawer and killed the aide. Again, the ensuing court case was a victory for the fundamentalist cause. Norris was acquitted on the grounds of self-defense, although it was clearly established that the aide had no weapon. However, legal victory and popular sympathy did not go hand in hand. Along with the caricature of fundamentalists as obscurantists, arose an image of fundamentalists as violent people, full of hostility and anger.

The last of the three events, which galvanized popular support against the fundamentalist movement, was the riot at Des Moines in 1929 – mentioned earlier in this chapter. Accusations of educational fraud against the school principal, the possibility of T. T. Shields having had an affair with his secretary, and pictures of the administrative offices torn to shambles were imprinted on people's imaginations. Shields himself withdrew to Canada where he continued to play a notable role in the fundamentalist movement. However, the Des Moines incident meant the end of his popular influence in the United States.

By the beginning of the 1930s, then, fundamentalism appeared to be in disarray everywhere. Scholarly studies sprang up which claimed that fundamentalism was the last gasp of a dying religious order that was quickly vanishing. Everywhere those of a more liberal persuasion celebrated theological victory. The truth was somewhat different. Instead of dying away, fundamentalists withdrew and began to build their own network of schools, churches, mission organizations, and radio programs. While the media pronounced defeat, conservative Protestants were quietly rearming. Thus, a partial listing of radio time in the United States in 1948 "indicated that over 1,600 programs were broadcast by fundamentalists each week."[7] While in 1954, the National Association of Christian Schools, an organization created after the Second World War to help provide elementary and secondary school education for children of fundamentalist parents, numbered 123 member schools located across the United States.[8]

Most of this activity went unnoticed by the mainline media and churches until some 40 years later, in the 1970s and 1980s, when fundamentalism once again began to influence public events, particularly in the United States. After appearing to the mainline media to be a dead religious movement, one could throw an egg out of a train window anywhere and hit a fundamentalist, to use an expression coined by H. L. Mencken at the height of the Scopes Monkey Trial.

To the consternation of mainline church leaders and much of the media, fundamentalism had not died but seemed to be on the ascendancy. In fact, the bestselling book on the *New York Times* book list in the 1970s was a fundamentalist book by an author Hal Lindsey. While updated and imaginative, *The Late, Great Planet Earth*[9] was really an unabashed restatement of the teachings of dispensational premillennialism, an integral part of North American fundamentalism.

Moreover, apart from Robert Schuller of Crystal Cathedral fame, the television and radio airwaves were dominated by those of a fundamentalist persuasion. It would have been H. L. Mencken's worst nightmare, had he lived. It was indeed the influential Canadian novelist Margaret Atwood's nightmare as she portrayed the United States dominated by fundamentalists who took over control of women's lives and who turned many into baby-making machines in her bestselling book *The Handmaid's Tale*.[10]

Even more unnerving, the fundamentalist movement seemed to have spread to other parts of the world and to other religious faiths. The media described the Iranian revolution as a fundamentalist takeover. Soon the word fundamentalist and Muslim became linked in popular perception. Jewish fundamentalists also made the news headlines, as well as Sikh and Hindu fundamentalist groups.

In 1980, Harvey Cox, who had predicted the growing secularization of the United States in the 1960s in his bestselling book *The Secular City*,[11] now had to revise his earlier prognostications. In a 1984 sequel, *Religion in the Secular City*,[12] Cox advanced the thesis that as modernity began to break apart, religious impulses again became potent forces. His two candidates for star billing in the coming decade were liberation theology and fundamentalism.

"GLOBAL" FUNDAMENTALISM

Protests against the use of this term *fundamentalist* to describe groups other than certain elements within North American Protestantism have arisen. Various reasons are advanced as to why the term "fundamentalist" is not adequate when applied to Islamic reform groups or to Sikh militants. The most compelling of these is the argument that the term arises out of a specific historical context and applies to a specific historical group.

Scholars also point to the inconsistency of the journalistic use. Often, for example, the nomenclature "fundamentalist" is used to describe Islamic and Sikh militants while conservative Jewish groups in the country of Israel are more commonly described by the term *ultraorthodox*. As Shupe and Hadden note:

This provides a significant clue to the biased political overtones of the concept. The bias is clearer still with press coverage of the Afghanistan rebel forces who successfully conducted guerrilla warfare against the Soviet army and the Soviet-controlled Afghanistan army. Portrayed as brave patriots fighting against the Soviet invaders, the mujahideen are Muslims engaged in a holy war (jihad). But because the Western world feels greater sympathy with these Muslims and their cause, the press has not characterized the mujahideen as fundamentalists.[13]

The use and misuse of the term "fundamentalist" by journalists has caused many observers of religion to discard this term, confining it to those strains of Protestantism which can be shown to be in historical continuity with the North American fundamentalist movement of the 1920s. For example, in a posting on the Political Islam Internet list, Steve Gotowicki, of the Department of Defense in the United States, notes:

...we here in the Department of Defense have increasingly turned away from the term Islamic fundamentalism to describe the political phenomenon we are witnessing in the region [i.e. the Middle East]. In its place we are prone to use the term Islamic radicalism or political Islam. This trend has also migrated to the Department of State, CIA, DIA, and other Washington area agencies.[14]

A large group of scholars, though, convinced by their research that there are important linkages on the ideological level between the various religious protest and reform groups throughout the world, have retained the term "fundamentalist," often adding the modifier "global" to describe this phenomenon. In order to be fair to non-Protestant and non-Western groups, they have attempted to empty the term of "its culture-specific and tradition-specific content and context,"[15] focusing on *four* features which are common to all fundamentalist movements, regardless of their cultural or religious context.

The first of these is a conviction that society is headed in the wrong direction. Thus scholars Jeffrey Hadden and Anson Shupe define fundamentalism, "as a proclamation of reclaimed authority of a sacred tradition which is to be reinstated as *an antidote for a society which has strayed from its cultural moorings.*" [emphasis mine][16] This conviction that society has strayed from its cultural and religious moorings and is heading in the wrong direction gives energy to organizations as diverse as Louis Farrakhan's Nation of Islam and a relatively new neo-evangelical Christian organization called the Promise Keepers.

Started in 1990 by former Colorado football coach, Bill McCartney, the Promise Keepers is a men's organization which is convinced that the vast bulk of social problems in the United States and Canada is due to the fact that men have abandoned their role as head of the family. In 1996, McCartney's organization had proved to be so popular that the annual revenue for Promise Keepers was $87 million and over 1.1 million men attended large group rallies in the United States.[17]

At these rallies men sing, shout, cry, listen to countless sermons, confess, share, and commit themselves to take up the mantle of leadership in their families once again. One of the best-known speakers is Tony Evans who outlines the strategy that men can take in reclaiming their role within their families.

The first thing to do is sit down with your wife and say something like this: "Honey, I've made a terrible mistake. I've given you my role. I gave up leading the family, and I forced you to take my place. Now I must reclaim that role." Don't misunderstand what I am saying here. [Evans continues] I am not suggesting that you ask your wife for your role back, I'm urging you to take it back.[18]

It would be unfair not to acknowledge the good which Promise Keepers has done in some marriages, or not to admit that men need opportunities to share together and support each other. Indeed, some African-American feminists are supportive of this organization, in that it forces men to take responsibility for their families, something which far too many men have failed to do. Moreover, the interracial emphasis within Promise Keepers has been praised by many as a redemptive and decidedly non-fundamentalist characteristic. However, the basic message, that society has lost its moorings due to the feminist movement and men's willingness to give up their leadership, is typical of global fundamentalism; as is the assurance that once men take back their role as head of the household, the problems of American society will begin to diminish and to disappear.

The second important feature of global fundamentalism is the conviction that this direction will lead to societal breakdown and that it *must* be reversed as soon as possible. In his book *Defenders of God*, Bruce Lawrence notes:

[Fundamentalism] is not one revolt but a series of revolts by those who uphold deep-seated religious values against what they perceive to be the shallow indeterminacy of modern ideologies. Rationalism or relativism, pluralism, or secularism – each undermines the Divine Transcendent, challenging his revelations, denying his prophets, ignoring his morally guided community. Without certitude the world is doomed. With it salvation (for some) is assured.[19]

In Protestantism, the typical expression of this conviction that society is on the brink of societal breakdown is the assertion that the end of the world is at hand. For example, the fundamentalist organization "Jews for Jesus" took out a full-page advertisement in *The Globe and Mail* following the Gulf War in which they intimated that Saddam Hussein was the antichrist. Titled

"Will Saddam Hussein Rise from the Dead?" the ad concluded, "if you're concerned about what has recently happened, is happening or may happen in the Middle East... If you don't believe in Y'shua (Jesus), or understand how prophecies can not only tell of him, but tell of the times we live in,... you'll want to read *Overture to Armageddon*."[20]

While this conviction that we are living in the last days is common to many different religious groups and cults, what differentiates the fundamentalists is that they believe that something can be done to reverse this trend, even if for only a very short while. In the early days of North American fundamentalism, individual conversion was the tactic of choice. North American fundamentalists reasoned that converted individuals would convert the larger society. In global fundamentalism, the stress usually begins with social conversion rather than individual conversion, making global fundamentalists much more overtly political than were their Protestant predecessors.

The third feature is the belief that only the *founding* religious tradition of the respective society in its *pure* form – the form advocated by the fundamentalist leader – can correct this problem and rescue society from inevitable destruction.

The perspective of Peter Marshall, the son of the former chaplain to the United States Senate, illustrates this aspect of global fundamentalism. After rebelling against his parents, Peter Marshall became a "born-again believer" in 1961. In time he entered the ministry and became the pastor of the East Dennis Community Church in Cape Cod. It was while Peter was speaking at a nearby chapel in Cape Cod that David Manuel, a New York editor, encountered him. Manuel describes that initial meeting,

Night had fallen on the small New England harbor, and the fishing boats rocked gently at harbor. Inside the nearby chapel, a gathering of some two hundred people were illuminated by electric candles which glowed softly against the wood paneling. The speaker came quickly to the heart of his message: "This nation was founded by God with a special calling. The people who first came here knew that they were being led here by the Lord Jesus Christ, to found a nation where men, women, and children were to live in obedience to Him... This was truly to be one nation under God."

The speaker paused. "The reason, I believe, that we Americans are in such trouble today is that we have forgotten this. We've rejected it. In fact, we've become quite cynical about it. We, as a people, have thrown away our Christian heritage."[21]

Convinced by what he heard and already believed, David Manuel cooperated with Peter Marshall in writing a book entitled *The Power and the Glory* in which they "creatively" traced out this heritage beginning with the voyage of Christopher Columbus and ending with George Washington's victories over the British armies. At the end of the book, the authors conclude that although the importance of Christian faith began to fade very early on in the formation of the country of the United States, leading to social degeneration, nonetheless no matter what had happened in the intervening generations, Americans would still be able to repent and to avail themselves of God's promises and "re-enter a covenant relationship with Him *as a nation*."[22]

Finally, the fourth key feature is the conviction that this religious medicine must be applied to all spheres of social life, not just to religious life or to religious institutions. As Marty and Appleby conclude in their arguments in support of the validity of the retention of the term "fundamentalism,"

Unlike many of their nonfundamentalist coreligionists, fundamentalists demand that the codes of behavior be applied comprehensively – not only to family life and interpersonal relations but to political organizations and international economies as well. Fundamentalists struggle for completeness because, as modern people, they have learned that traditional life based in the home, school, village, or tribe is not sufficient to ward off the invasive, colonizing Other. The religious community must therefore reject artificial distinctions between "private" and "public" realms – distinctions too easily accepted by conservative or orthodox believers who prefer to "live and let live." Fundamentalists know that life itself depends on victory over the enemy in a war for the control not only of resources but of ideas...[23]

THE SACRED-SECULAR DUALISM

This rejection of the dualism of sacred and secular (or as Marty prefers – private and public), which was hammered out in the Middle Ages with the establishment of the clergy as distinct from the laity and found its full fruition in the rise of modern science in the 1600s and 1700s, also helps to explain the paradoxical relationship which global fundamentalism has with the modern world. Many observers of contemporary religion have labeled fundamentalism as a premodern movement which seeks to move the societal clock back in time. Such an assertion is deceptive, however; groups such as the Amish in North America, who reject most modern amenities, are not part of the global fundamentalist movement. Moreover, often global fundamentalist groups are "armed" with the very latest in modern technology, be it weaponry of the body, as in the case of many fundamentalist Muslim groups or, weaponry of the mind, as in the case of North America fundamentalists who control the lion's share of religiously orientated radio and television shows.

Fundamentalists have such a paradoxical relationship with modernity because it is the modern split between the world of the sacred and that of the secular which brings such groups into being. Thus, in many countries where the Muslim faith predominates, fundamentalist Muslim groups do not come into being until the country is westernized or modernized. Bruce Lawrence notes,

...fundamentalists do not deny or disregard modernity; they protest as moderns against the heresies of the modern age. In the Salman Rushdie affair they confronted freedom of speech with loyalty to age-old cultural norms. To the makers of The Last Temptation of Christ *they posed blasphemy as a higher standard than artistic freedom, while in Israel they favored occupying* all *the Land of Israel* (Eretz Yisrael) *rather than pursuing pragmatic policies dictated by the secular state of Israel.*[24]

This rejection of the dualistic split between the sacred and the secular explains why, in North America, fundamentalists of the past such as William

Jennings Bryan protested against the infiltration of the teaching of evolution. Many Christians made their peace with Darwin's theories through the assertion that the creation chapters at the beginning of the Book of Genesis were concerned with the question of *why* the world was created, the *how* was a secondary question which could be left up to science; fundamentalists rejected both the compromise and evolutionary theory – the Bible proclaimed not only *why* but also *how*.[25]

Moreover, this rejection of the sacred-secular split is also why fundamentalists of the present such as Jerry Falwell, Pat Robertson, and others in the United States and, on a much reduced scale, Ken Campbell in Canada are so active politically. On one level, it would make sense for such leaders to rejoice that North America is on the slippery road to hell since, according to premillennialism, this is a necessary step before the rapture and the end. However, the call of a Christian civilization, romanticized though it may be, also speaks to fundamentalist hearts, whispering seductively whenever the time seems right.

Indeed, so strong is the call for a unified society in which religion interpenetrates political as well as personal life, that Falwell was willing to abandon the doctrine of separatism in order to cooperate with non-fundamentalists in the reformation of American society. The doctrine of separatism – the belief that fundamentalists must not cooperate with non-fundamentalists – became one of the key definitions of the true fundamentalist. Indeed, prominent fundamentalist leaders censured even the conservative evangelist Billy Graham when, in the 1950s, he began to accept the involvement of mainline church leaders in his famous crusades. By the 1980s, though, what was unacceptable for Billy Graham's evangelistic crusades became acceptable for Jerry's Falwell's political activities.

THE POLITICIZATION OF NORTH AMERICAN FUNDAMENTALISM

This emphasis upon a god-fearing society, then, offers important clues as to why North American fundamentalists have been able to modify substantially the premillennialist thrust which scholars such as Ernest Sandeen insist is the chief definition of fundamentalism.[26] The Protestant world view before the growth of fundamentalism was marked by an emphasis on *postmillennialism* not *premillennialism*, particularly in the United States.

Church leaders such as Jonathan Edwards, a colonial preacher and theologian, taught that the return of Christ would take place *after* the millennium which was usually interpreted literally as a 1,000 year period of peace and prosperity. Although evil would not be totally eliminated during the millennium it would be reduced to a minimum and with the return of Christ would eventually disappear.

In contrast, premillennialists claim that there will be a 1000-year rule of peace on the earth *but* that this will not come about through human efforts. Instead, they picture a time of growing societal degeneration marked by famine, plague, war, and widespread apostasy. This period of tribulation, often depicted as lasting seven years, will only be ended by the spectacular intervention of Christ who will return to earth, defeat Satan, conquer evil, and inaugurate the millennium.

The important point to remember, though, is that while the United States was never the Christian nation which some fundamentalists like to portray that it was, most American Protestants up to the time of the rise of fundamentalism believed that the United States was the embodiment of what God was doing in history. An optimistic spirit prevailed in both the explicitly religious form of postmillennialism and in the more secularized form of belief in inevitable progress.

Indeed, it is possible to claim that it was not until the assassinations of John Kennedy, Martin Luther King Jr., Robert Kennedy, the Vietnam War, and Watergate that this optimistic spirit began to recede in the United States like an Atlantic tide along the Bay of Fundy. Even then, the actor-turned-president, Ronald Reagan, using the threat of the evil empire of the Soviet Union, was able to resurrect the crusade to incarnate God's kingdom until the oil crisis and the threat of strangulation by pollution finally killed it.

Following the Civil War, long before this massive retreat by American society from utopian visions, a group of Christians began to migrate from postmillennial optimism to a more somber premillennial interpretation of history. In premillennialism, as has been stated, things are not getting better and better but worse and worse. The end is right around the corner and it will be an ugly end at that. By the time fundamentalism made its appearance in the late 1920s, premillennialism had become an accepted teaching. So much so that Ernest Sandeen in his work on fundamentalism insists that along with biblical literalism, the other defining feature of North American fundamentalism is a belief in premillennialism.[27]

The reasons for this migration from postmillennial optimism to premillennial pessimism are varied and fascinating. The most incisive study which tries to tease out the reasons for this change is Douglas Frank's book *Less Than Conquerors: How Evangelicals Entered the Twentieth Century*.[28] Frank argues that, faced with a diminution of political and social power due to urbanization, immigration of large numbers of Roman Catholics, consumerism, Darwinism, and biblical criticism, a segment of the evangelical community (the proto-fundamentalists) grasped the final power which was left to them – the knowledge of when the end of the world was to come.

Before this shift could happen, though, a prior shift also had to take place. In postmillennialism, it was easy to know when the end would come. It would follow on the heels of a 1,000-year reign of peace and prosperity. If the millennium was to happen following the return of Christ, though, then suddenly the rules of the game shifted. Accordingly, proto-fundamentalists began to look for signs of decline within Western civilization rather than indications of progress. These they found in the spread of Darwinism, the growth of the Roman Catholic church, changing social roles for women, and the infiltration of liberal biblical criticism in many theological schools. Moreover, to make things even more acceptable, the concept of the rapture – the taking up of all Christians immediately before the tribulations that were to accompany the end of the world – made an appearance. Now the suitably enlightened believer could safely shift from postmillennial thinking to a premillennial perspective secure in the knowledge that a) he or she would not have to face the troubles of the end times (being raptured safely to heaven before it all started) and b) he or she had access to the power of knowing a secret which made any weapon of the day look like a toy pop gun.

The intriguing question, then, becomes why the resurgence of North American fundamentalism in the 1970s and 1980s modified this premillennial pessimism moving in a more postmillennial direction. Typical of the confusion this caused, both within and outside fundamentalist circles, was a question Pat Robertson was confronted with at a pastors' luncheon in Concord, New Hampshire, as he was deciding whether to run for president of the United States. "Wait a minute," one of the ministers said. "The next event on the eschatological clock is the return of Christ. Things in society should get *worse* rather than better. If Christians worked to turn our nation around, that would be a humanistic effort and delay Christ's return."[29]

What this minister was unaware of, though, is that along with premillennialism there existed an older and somewhat contradictory emphasis within fundamentalism on the establishment of a Christian civilization in Canada and the United States. When the media and the mainstream churches rejected fundamentalism, causing fundamentalists to focus their attention elsewhere, premillennialism reigned supreme within fundamentalist circles. However, when fundamentalist numbers began to grow and mainstream church numbers decline, the emphasis upon the priority of a Christian civilization again could come to the fore.

That this shift signaled a return to a more optimistic viewpoint did not seem to faze Pat Robertson or Jerry Falwell. The end of the world was still to come but, paradoxically, it would be ushered in not by an inexorable slide towards godlessness and moral anarchy but by a brief surge in Christian commitment and political influence.

It is this focus upon a Christian civilization which compelled me to add a chapter on fundamentalism in this book on new movements in religion and spirituality. Besides the fact that fundamentalism is new (an assertion which has garnered me much criticism from fundamentalist supporters who like to believe that they are simply preserving the faith), it is also an important aspect of the shift from a modernistic to a postmodernistic society.

PROBLEMS WITH THE
FUNDAMENTALISM VISION

The problem with this fundamentalist vision of a Christian/Jewish/Hindu/ Muslim society is not the rejection of the sacred-secular dualism as many people claim. "The church has no business in politics" is what one often hears and there is some truth in such a claim if by the word *church* one refers to the institutional face of Christian faith. However, if by the word *church* one means people who claim a religious faith, then it cannot be a bad thing that the individual seeks to integrate his or her faith.

I remember a ministerial meeting where Flora MacDonald was our guest. Flora served at that time as Minister of Foreign Affairs for Canada and was the darling of the American press for her involvement in helping to spirit several Americans out of Iran. Her topic was the role of the church in politics and she made a plea for participation by the Christian community in the political forum as she lamented the lack of involvement by church groups. At that, one of the ministers put her on the spot. "What are you doing in your life to integrate your faith with your political action?" he asked. Surprised by the question, she offered no reply.

Rejection of the compartmentalization caused by the dualism of sacred-secular is not the problem with the fundamentalist social agenda. The problem is that a new dualism creeps in which is just as devastating as the former. Instead of a sacred-secular dualism, fundamentalists succumb to a good-bad dualism.

Of course, I should not be too harsh on the fundamentalist movement, particularly in its Christian form, since Christianity has as a whole suffered from the dualism of good versus bad, of God versus Satan. In the Old Testament, Satan appears as a servant of God to test the faith of God's people. By the time that the New Testament books were written, Satan is depicted as almost equal with God and dualism inherited from the Zoroastrian faith had begun to settle in like a maritime fog on a cool autumn evening.

In the Roman Catholic church, this dualistic emphasis is held in check by an emphasis on the incarnation and the original goodness of the created order. In Protestantism, it is held in check by an emphasis on the distinction between the church and the Kingdom of God. In opposition to Roman Catholicism which equated the Kingdom (activity/reign) of God with the church, Protestants realized that the reign and activity of God, through the Spirit, was much wider and greater than the parameters of the visible church. Unfortunately, fundamentalists adopt a Roman Catholic view whereby they equate the visible church with the Kingdom of God but reject the Catholic emphasis on the importance of the incarnation. In fundamentalist churches, the Mass is not seen as the body and blood of Christ, a very graphic reminder that God has interpenetrated this physical world of ours, but as symbols of a spiritual rather than a physical presence. This means that fundamentalists divide the world into two camps, the good side, to which they and their allies belong, and the bad side, to which all others belong. This allows fundamentalists to take a very cavalier attitude towards the preservation of life and of the world itself. In an age of nuclear technology when we humans are able to destroy life quickly through atomic warfare or more slowly through environmental pollution, this view is disturbing.

Consequently, fundamentalism cannot provide the social vision for the religious and spiritual ferment that is occurring around us. Dualisms of whatever sort do not adequately deal with the complexities of life because they divide and wall off. Only a monism, an emphasis on the oneness of all life, can be open to the pluralism within our world. Monism and pluralism complement each other, like two sides of the same coin. Fundamentalists cannot affirm the oneness of life nor can they allow for pluralistic expressions of spirituality. There is one right way and one wrong way, and to their last breath, the fundamentalist will battle for that right way.

It is my conviction that in the future a healthy spirituality and, consequently, a positive religious, social vision must rise above dualistic thinking to embrace a more holistic view of life. As the little poem puts it, "Rebel, heretic, thing to flout. He drew a circle to keep me out. But love and I had wit to win, we drew a circle that brought him in." Fundamentalism refuses to draw the inclusive circle, which, unfortunately, is also the problem with the next socio-religious movement that I will examine – Latin American liberation theology.

8

LIBERATION THEOLOGY

A short news notice in a recent edition of the local newspaper captured my attention. Thirty years after he had died in the country of Bolivia, the bones of Che Guevara, the Cuban revolutionary fighter, were returned to his home country of Cuba. Once there, they were reburied with all the pomp and circumstance befitting a hero of the revolution. I was living in Bolivia at the time of Che's death. My parents were working there as missionaries and my father's assignment at that time was to serve as director of a Baptist mission radio station located in the capital city of La Paz.

La Paz is a scenic city. Perched some 13,000 feet above sea level in a small bowl which offers protection from the Altiplano winds, the city is dominated by the spectacular vista of Illimani in the distance, surely one of the most majestic mountains in the world. Above the rim of the city sits the Altiplano, like a tan brown, weather-beaten saddle stretched between the western cordillera on the one side and the eastern cordillera on the other. In Bolivia, the Andean mountains split for a while until they reunite further south to form the backbone of the countries of Chile and Argentina, ending finally in the Tierra del Fuego, at the southern tip of South America.

Life in La Paz was vastly different from Toronto, where we had come from, but like children everywhere I soon adjusted. After all, having lived in the Bolivian tin-mining town of Oruro before my family's sojourn in Toronto, I considered myself more Bolivian than Canadian. One of the things I was used to as an adopted Bolivian was the omnipresence of the Bolivian armed forces. As kids, we would joke that in Bolivia you didn't look forward to snow days (it snowed only twice that I can remember during my time there), but rather to revolution days. At least once a year and sometimes more often there would be an attempted coup. Sometimes this would result in a *golpe* in which the government would be overthrown and a new one instituted in its place. The presence of soldiers and tanks did not bother me, except on October 8, 1967, when a squad of machine-gun armed soldiers took over the radio station where Dad was working. We lived nearby in a house on Calle Argentina and the radio station was one of my frequent haunts. Even then I was not overly worried about the presence of the soldiers until I saw my father's face and realized that he was deeply concerned.

I found out later that Che had been killed that day, according to Fidel Castro – a day before, according to Bolivian reports.[1] A group of Bolivian soldiers, aided by the United States military, slowly tightened the noose around Che and his guerillas. On October 7, 1967, Che was taken captive, badly wounded; his submachine gun jammed from a stray bullet. In his belongings was a diary which he had kept during his time in Bolivia; a country which because of its grinding poverty, rugged geographical terrain, and high percentage of Indian population was considered ripe for a communist revolution. In the diary, Che mentioned listening to two radio stations for news and information. The first station was Radio Habana; the second was our mission radio station, Radio La Cruz del Sur (the Southern Cross Radio) named after one of the most prominent constellations in the southern hemisphere.

The Bolivian military must have speculated that coded messages were being sent to Che by radio, prompting their swift action. The radio station was given back soon after due to protests and in spite of suspicions that one of the announcers harbored leftist tendencies and therefore took an antigovernment position. I remember puzzling over this at the time, because my Baptist heritage had convinced me that religion and politics were completely divorced from each other. If there was a connection between the two, I knew that it would be in support of a Western-style democratic government, not a Cuban-style socialist one. How could a radio announcer, who was presumably also an active Christian, be in sympathy with leftist guerillas such as Che?

Aside from this short takeover of the radio station I can remember only one other instance when it dawned on me that there was a new ferment in Bolivia, although I only vaguely dreamed that it was connected with the church. I was out walking when I turned a corner and came to an intersection ringed with soldiers. In the middle of the intersection was a pile of rectangular stones torn up from the cobblestone street. I quickly realized that it would be best to get out of there and was about to do so when I noticed an Aymara woman being beaten by a soldier. She was wailing more loudly than the situation warranted, as I remember, but somehow the male chivalry that I had been raised with reared its head. I started walking toward the soldier, yelling at him to stop hitting the woman. I don't remember what happened next, but suddenly university students were running here and there pursued by soldiers shooting tear gas canisters. I ran as fast as I could and hid behind a mud wall, crouching silently until the excitement had died down. Later, when everything was quiet, I crept over to an empty tear gas canister that was lying on the roadside. I picked it up and took a sniff, realizing right away why tear gas was such an effective way of dispersing an angry crowd; my eyes stung and I gasped for breath.

The same sense of surprise which I experienced upon finding out that a Christian radio announcer at our mission station had leftist sympathies came over me ten years later, while I was shopping at the United Church bookstore in Toronto. I was enrolled in a political science course at university and wanted to write my assignment on some aspect of Christianity and politics; by now I had learned the two were not as distinct as I had previously thought. However, I retained my belief that Christianity and Western-style, liberal democratic government went hand in hand and so, when I chanced upon a book, *Marx and the Bible*, I stopped short. "That's curious," I mused to myself. "What does Marxism have to do with the Bible?" I looked at the cover, which depicted a scene I had witnessed many times in Bolivia. A *campesino* (peasant) was patching a white porcelain enamel bowl, ringed with a blue band. In the United States and Canada we would have thrown the bowl out long ago, but in Bolivia it would be patched and repatched until it finally fell apart and was of no use. On impulse I bought the book. So began a fascinating journey into a theological and religious movement which had been going on all around me in Latin America, but of which I was almost completely unaware.

THE GROWTH OF
LIBERATION THEOLOGY

Liberation theology, the subject of José Porfirio Miranda's book *Marx and the Bible: a Critique of the Philosophy of Oppression*, is the second contemporary Christian movement which has attempted to present a religiously inspired social/political vision, not only for Latin America but for North America as well. It was the first Christian theological movement to germinate in Latin America, even though its ideological roots lie in Europe. A group of Roman Catholic priests and sisters who were born in Latin America traveled to Europe for their theological studies. While studying in Europe, they were exposed to fresh currents of thought sweeping through the Roman church because of the religious renewal that culminated in Vatican II. Part of that renewal was a growing dialogue between Christians and Marxists. *Rapprochement* would be too intense a word to apply to this dialogue but there was recognition among many clergy that Marxism was not the bogeyman that they had been led to believe.

These priests and sisters carried their new vision back to Latin America where they were met by a church hierarchy firmly allied with the ruling powers, but slowly becoming receptive to change. Although liberationist writers and popular films such as *The Mission* would lead one to believe otherwise, Latin American Catholicism was *not* a people's movement. In part this was due to historical reasons – as Catholicism and colonialism went hand in hand. But it was also due to the traditional viewpoint of the Roman Catholic church which taught that the secular state had a subservient role in regard to the church. The church wielded two swords, the swords of spiritual power and political power, while the state wielded only the sword of political power. The spirit of Vatican II, however, was felt even in conservative Latin American Catholicism. Although it took a few years to cross the Atlantic, it finally did, manifesting itself at a bishop's conference convened at Medellín, Colombia in 1968 – a conference that has often been hailed as the Vatican II of Latin America.

Several historic events besides Vatican II helped prepare the way for the perspectives that surfaced at the Medellín conference. The most important of these was the defeat of the Baptista regime in Cuba by Fidel Castro, Che Guevara, and others in 1959. The Catholic church in Cuba, as elsewhere in Latin America, had been identified with the ruling elite and not the poor. This, as well as the atheism which is part of Marxism, prompted Castro to condemn large segments of the church and to curtail its role within Cuban society. Although many in the church hierarchy saw this as a great offense, others, such as the Colombian priest Camillio Torres, agreed that the Roman Catholic church had been part of the problem of poverty. A member of a prosperous family in Colombia, Torres' social and political views shifted dramatically while studying at Louvain. In time, he was led by his revolutionary commitments to renounce the priesthood and to fight as a guerilla against the Colombian government. He explained why he left the priesthood stating,

I have ceased to say Mass [in order] to practice love for people in temporal, economic and social spheres. When people have nothing against me, when they have carried out the revolution, then I will return to offering Mass, God willing... I think that in this way I follow Christ's injunction... "Leave thy gifts upon the altar and go first to be reconciled to thy brothers."[2]

In February 1966 he was shot and killed by government troops.

Torres' controversial example, along with the progressive atmosphere of the Medellín conference, provided the soil in which liberationism could flower. In 1971 the Peruvian priest Gustavo Gutiérrez published a book that gave prominence to this ferment within the Roman Catholic church. The book *A Theology of Liberation*[3] was never meant to be the definitive text of the liberation movement, which many English observers have taken it to be. Even in the English translation of 1973, it was clear that the book was conceived of as only one attempt to apply the insights of the Bible, as refracted through theological training in Europe, to the concrete realities of life among the poor in Latin America. It was soon followed by a host of other books on the theme of liberation, indicating that something was afoot far bigger than one individual.

The theological systemization of this new movement was left to the Jesuit, Juan Luis Segundo. While he worked as director of the Peter Faber Pastoral Center in Montevideo, Uruguay, Segundo wrote a five-volume series under the title *A Theology for Artisans of a New Humanity*. Later, he condensed his thoughts into a one-volume work entitled *The Theology of Liberation*.[4] A bewildering array of articles and books followed and liberation theology became well known not only in Latin America but also in North America where connections were made with Black and feminist theology.

Soon the spread of the movement, its use of Marxist class analysis, its support for the poor against the rich, and the growth of small Protestant-like prayer groups known as *communidades de base* (base communities) prompted a counterattack by the Roman Catholic hierarchy, both in Latin America and in the Vatican. In Latin America, the counterattack was led by Bishop Lopez Trujillo who was elected secretary-general of the Latin American Bishops Conference in 1972. Once in that position, Trujillo was determined to minimize the influence of the liberationists. Liberation theologians either were replaced by more moderate church leaders, or the power of their positions was severely diminished. At the same time, though, the liberationists were establishing partnerships with Catholic and Protestant leaders and academics in North America and Europe. Important meetings to discuss liberationist concerns were held at Mexico City and Detroit in August 1975; the next year, in Dar-es-Salaam, the Ecumenical Association of Third World Theologians was formed to further the liberationist cause.

The stage was set for a civil war within the Roman Catholic church in Central and Latin America when preparations began for the Puebla Conference in 1976. The stated purpose of this conference was to study and evaluate the ecclesiastical process begun at Medellín, but the real purpose, according to liberationists and their supporters, was to demolish the influence of the liberation movement. Postponed by the death of two popes, Paul VI and John Paul I, the meetings finally began on January 28, 1979. The new pope, John Paul II, received a tumultuous welcome in Mexico City and spoke at the Shrine of Our Lady of Guadalupe before traveling to the city of Puebla, some 75 miles away, to open the conference. Extensive press coverage by both the Mexican and international press seemed to guarantee a religious civil war. It was not to be.

Although Pope John Paul II distanced himself from the Marxist overtones of liberationism and its openness to violence in the pursuit of economic equality, he did not condemn the movement. Moreover, the emphasis on a "preferential option for the poor" which surfaced at Medellín and had elicited so much controversy was reaffirmed.

With renewed hope in the vivifying power of the Spirit, we are going to take up once again the position of the Second General Conference of the Latin American episcopate in Medellín, which adopted a clear and prophetic option expressing preference for, and solidarity with, the poor. We do this despite the distortions and interpretations of some, who vitiate the spirit of Medellín, and despite the disregard and even hostility of others. We affirm the need for conversion on the part of the whole Church to a preferential option for the poor; an option aimed at their integral liberation.[5]

The peace at Puebla, however, masked deep changes in the Roman Catholic church in Latin America and elsewhere. Everywhere religion's influence on political life seemed to be on the rise; the Iranian Muslim takeover, the activities of the new right in the United States, and in Latin America, the once conservative and traditional Roman Catholic church reaffirmed the struggle for liberation on behalf of the poor. The religious genie had been let loose.

Six months after the Puebla meetings, the Sandanista revolution in Nicaragua, which toppled the Somoza regime, was remarkable for the high level of participation of priests and sisters. One of the best known of these Nicaraguan liberationist priests was Ernesto Cardenal. Inspired by the example of the Trappist monk Thomas Merton, Cardenal started a contemplative community in Nicaragua on the island of Solentiname in Lake Nicaragua. This contemplative community, however, was not a traditional one, but rather a conscience-raising community. The Bible was read through liberationist spectacles. Cardenal himself became enchanted with Cuba and with Fidel's socialist movement. Soon he would claim that "Mary was a revolutionary and a Communist."[6] In time, the National Guard raided and destroyed the community because they claimed that it had become a guerilla base. Cardenal went into exile in nearby Costa Rica until he returned in triumph, with the overthrow of Somoza, to assume the position of Minister of Culture in the new government.

Cardenal's political activities proved to be too much for John Paul II. In a visit to Nicaragua in March 1983, he publicly scolded Cardenal in front of the television cameras, urging him to regularize his relations with the Roman hierarchy. At an open air Mass that followed, though, Cardenal got his revenge as the pope was preempted by a group of Nicaraguan mothers who demanded that he condemn the United States-led *contras*. The Mass ended with a visibly shaken pope and the singing of the Sandanista hymn denouncing the Yankees as the enemy of the human race.

In spite of the Sandanista's success, however, liberationist opponents found support in the election of Cardinal Ratzinger as chair of the Vatican Congregation for the Doctrine of Faith, a watchdog agency of Roman Catholic doctrine, historically connected with the Sacred Congregation of the Universal Inquisition, founded in 1542. With Joseph Ratzinger's appointment in 1982, the controversy intensified that had been brewing between a Brazilian priest, Leonardo Boff, supported by several prominent Brazilian bishops, and the Vatican. In 1985, Boff was sent a letter indicating that a book on church structures that he had written contained several essential errors of doctrine. In April of that year, Boff's Franciscan superiors were asked to impose a period of silence on him, during which he was not to preach, give interviews or publish. Boff accepted the church discipline, which was lifted a year later, on March 29, 1986.

The conflict between Ratzinger (and by extension Pope John Paul II himself) and various liberation theologians, as well as changing political conditions in Latin America, toned down the early Marxist rhetoric of the liberationists. Instead of a blanket condemnation of Western-style democracies, there is now openness to democratic change. Moreover, Gustavo Gutiérrez, the dean of liberation theology, increasingly has emphasized the importance of spirituality in his own life and writings, rather than overt political action.

THE ECONOMIC AND POLITICAL CONTEXT

Key to understanding the liberationist movement is the importance of the political and social context out of which it arose. Here is an important difference between the social vision put forward by fundamentalism and that advanced by liberation theology. Fundamentalism is "one size fits all." There is one way, one book, one faith, one vision of culture, which is the right way for all, regardless of gender, racial, religious or economic differences. Fundamentalists are blithely blind to the social, political, and historical context that fashioned their movement. In contrast, liberationists take their social and historical context seriously.

It was the context of widespread and brutal poverty that coaxed the liberation movement into being. During the 1960s and 1970s in Latin America, it became clear to the Two-Thirds Nations that the First World would not only *not* help them economically but was actively engaged in keeping them dependent. Moreover, though often rich in natural resources, the wealth garnered from their exploitation did not reach the typical Latin American; instead, it was divided between a small indigenous upper class, and foreign, typically American interests. As Aharon Sapsezian noted in the 1970s,

Poverty in Latin America is a blatant, chronic and worsening reality! Not only do large sectors of society lack access to the basic necessities normally associated with human dignity; they are even denied the freedom to take any initiatives aimed at alleviating their plight by means of structural changes conducive to a less iniquitous distribution of material resources.[7]

Today, the dependency theory which postulated a continually increasing reliance of the Two-Thirds countries upon First World countries has had to be modified somewhat due to the economic betterment of certain Far Eastern countries and the success of the Arab nations in receiving a higher price for their natural resource of oil. Such has not been the case in Latin America. An attempt by Jamaica to form a cartel in order to obtain a fair price for bauxite, an essential ingredient in the making of aluminum, failed. The government of Jamaica was destabilized in the process and the economy worsened.

In part, continued poverty in this area has been due to the proximity of Central America, Latin America, and the Caribbean nations to the United States. Latin America has been colonized twice. The first time by the Spanish and the Portuguese, who came grasping gold with one hand and offering God with the other. Later, after the decay of the Spanish and Portuguese empires, this area was claimed by the United States as its colonial space, as indicated by the Monroe Doctrine issued in 1823. The Americans repeated the Spanish and Portuguese pattern, bringing Christ in one hand and taking capital with the other. In such circumstances, it is natural that the stress of the radical young priests and sisters would be first on economic and political liberation and second on spiritual liberation.

This emphasis on the social and historical context as a key determinant in the formulation of the church's theology has meant that the export of the theology of liberation has resulted in different liberationist movements, depending upon the situation in which they arise. In Asia, the presence of Eastern faiths has given rise to a fascinating interreligious dialogue. In Africa, the theme of racial liberation has been predominant. In North America, along with Black and Hispanic liberationists, feminists and homosexuals have taken up the liberationist method and message.

The pluralism engendered by the focus on *context*, in contrast to the previous belief that a good theology was applicable anywhere with only a few minor variations in emphasis and tone, has not been without problems. At a seminar held at Princeton University with Robert McAfee Brown, one of the principal interpreters of this new movement to the English-speaking world, I participated in a round-table discussion. The group was composed of several supporters of Black liberation, a majority of supporters of feminist liberation, and me as the lone advocate of liberation theology proper. At one point in the discussion I advanced the argument that the large influx of women into the work force in North American society had the net effect of driving up the price of goods, particularly housing, and further impoverishing the lower class. I soon learned that the priority of eradicating poverty as advocated by the liberationists, was very different than the priority of gender balance advocated by the feminists! I still maintain that a rich woman and a poor woman have less in common that a poor woman and a poor man, but I have become far more nuanced and sensitive in my comments than when I was an eager convert to the liberationists' cause.

The context of poverty which the liberationists faced upon returning to Central and South America from studies in Europe not only forced them to take seriously the context in which one theologizes, but also to put *praxis* (action) ahead of theological reflection. Rather than *orthodoxy* (correct doctrine), the liberationists called for *orthopraxy* (correct action). Theology was something that was done after the sun had set; during the day, one worked at bettering life for the poor.

A Brazilian Baptist living in Canada once told me that he had visited several of the small base communities, which have become an integral part of the liberationist movement, when he returned home to Brazil for a family visit. He described them as typical Protestant prayer meetings. First, there was some singing, then prayer, then study of the Bible. The difference, said Joao Keidann, was that at the end of the meeting rather than going home to bed and forgetting what they had learned, the participants applied the biblical passage to their everyday situation and put the lessons into action.

Jon Sobrino, a Jesuit priest and prominent liberationist, claims that it is this emphasis on *praxis* which has been liberation theology's most important contribution. He argues, "the most original and fruitful features of the theology of liberation, more than any concrete thematic content, remain its mode of understanding the business of theology as a theoretical reflection on praxis..."[8] The task, as liberationists understand it, is to change the world rather than understand it; or better put, to seek to understand the gospel mandate through changing society.

A focus on praxis (action) goes hand in hand with the most controversial element of liberationism: the use of Marxist class analysis and the willingness to support violent means to achieve desired ends. According to Juan Tamayo,

European theology has its ties with the First Enlightenment (Kant) and its interest centers on freeing reason from authoritarianism and dogmatism or in other words, the deliverance of an enslaved subjectivity.

By contrast, the theology of liberation seeks to respond to the challenge of the Second Enlightenment (Marx). In the eyes of this theology, the liberating function of knowledge becomes concretized in the transformation of reality, and thereby recovers the threatened meaning of faith.[9]

Critics of liberation theology claimed, however, that rather than simply responding to Marxism, liberationists had been taken over by Marxism. In an attack on liberation theology, the American writer Ronald Nash insisted that, "socialism can provide neither bread nor freedom.... Economic systems that decrease or discourage production can never succeed in eliminating poverty; they can only make it worse."[10]

The problem with such criticisms is that they do not take into account the selective use of Marxism by the liberationists. Indeed, at times, it seems as if almost anything to do with a commitment to the poor is labeled as Marxist by North Americans. I once had to speak at the wealthiest Baptist church in Toronto. The topic was whether the poor threatened the stability of the world economy or not. My opponent was an advocate of lifeboat ethics. "The poor," he stated, "are breeding like rats, threatening to capsize the good ship Earth." To clinch his argument, he added rhetorically, sweeping his hand around the elegant room we were meeting in, "we certainly wouldn't want our beautiful Yorkminster Park overrun by rats would we?" I gamely argued back that the Two-Thirds World consumed less food, less energy, less wealth and that what the good ship Earth was threatened by was not the inhabitants of the Two-Thirds World but the inhabitants of the First World who were using more than their share of the Earth's resources. At the end of the debate, which I lost, a member of the audience thanked me for giving voice to Marxist liberation theology!

Liberationists, stung by the criticism of having sold out to atheistic Marxism, have responded by claiming that their use of Marxist thought has been selective. As Enrique Dussel puts it, liberation theology adopted a Marxism, which was "compatible with a Christian faith received from the prophets, from Jesus, and from church tradition..." He adds, "Anything like a Stalinist dogmatism, the economicism of the manuals, or 'philosophical' Marxism is altogether foreign to liberation theology."[11]

The elements within Marxism that the liberationists used were Marx's stress on economic and historical determinism. The Marxists blamed the class struggle and the heavy hand of capitalist oppression for keeping the poor impoverished; the answer to this situation was liberation through revolution. Not all liberationists advocated the use of violence. Bishop Dom Helder Câmara, whose diocese includes the city of Recife at the mouth of the Amazon, for example, was an outspoken opponent of the use of violence, even in the pursuit of lofty aims.

Most liberationists, however, claimed that Thomas Aquinas' just war theory applied to their situation. They were not practicing violence, they claimed, but counterviolence. Violence was already institutionalized in ways that left many malnourished and devoid of hope, both physically and spiritually. The just war theory meant that the use of force to oppose this prior violence was morally justifiable.

An apocryphal story which may have some factual basis illustrates the liberationist's position. Daniel Berrigan, the American priest who was active in the opposition movement to the Vietnam War and later prominent in the call for nuclear disarmament, is said to have confronted Ernest Cardenal and commented to him, "All ideologies must fall in the face of the death of even one innocent child." Cardenal never missed a beat as he replied, "Yes, even the ideology of pacifism."

Use of violence on behalf of the poor was given church blessing not only by Aquinas' just war theory, according to liberationists, but also by the call for a "preferential option for the poor," first articulated at the bishops' conference in Medellín and reaffirmed at the conference in Puebla. At first, this phrase simply meant that the religious community in South and Central America should make the eradication of poverty an important issue. One way in which this was to be done was for priests and sisters to embrace material poverty in order to walk with the people to whom they were ministering.

The Bible also was read in light of this option for the poor and it became clear to many that European and North American exegetes had habitually misread certain biblical passages and ignored others. The version of the first beatitude contained in the Book of Matthew, for instance, which reads "blessed are the poor in spirit for theirs is the kingdom of heaven," was elevated; Luke's version which simply reads "blessed are the poor" was ignored or reinterpreted in light of Matthew's emphasis on humility. The liberationists claimed that Luke's rendition, in which the physically poor are blessed while woe is pronounced upon the materially rich, is closer to Christ's original intent.

In time, the preferential option for the poor began to take on overtones which the bishops never meant. Hugo Assmann, for instance, used the phrase the "epistemological privilege of the poor" to emphasize his belief that the poor are both open to God's call in a way the rich are not and that the poor are the locus for God's saving activity. Violence against the rich was thus justified. The rich stood in the way of what God was doing with the poor, the future belonged to the poor, to the proletariat, and the fastest way to that future was to get rid of the rich.

BASE COMMUNITIES

One of the most interesting aspects of liberation theology, from a ministerial perspective, is the establishment and growth of the small house churches known as "base communities." I have already alluded to comments made by a Brazilian Baptist in which he claimed that these base communities were similar to Protestant prayer meetings. Others have made even grander claims, stating that the liberation movement is a new Protestant Reformation in which an oppressed wing of the Roman church is throwing off the tyrannical domination of Rome. Such rhetoric, however, says more about the observer than about the base communities.

There is a shift taking place in the Roman Catholic church worldwide, from being a clergy-dominated church to one in which the laity have a greater say. I remember in Moncton, New Brunswick, the priest at one of the largest Roman Catholic churches complained to me that lay control had gotten out of hand. Originally, this priest was a strong advocate of more lay involvement, but increasingly he came to see the price tag involved. It is as part of this growing involvement and control by the laity that the Latin American base communities should be situated. As the Roman Catholic church continues to fail to produce enough clergy, sisters and lay people have stepped in to fill the gap. Logically as these sisters and lay leaders have carried the pastoral load, they have felt justified in calling some of the shots. In Latin America, for example, many small communities had no priest; a peripatetic priest would come to say the Mass periodically, but control of the church community fell to lay leaders.

At one time, it seemed, both to the Roman hierarchy as well as to the world at large, that the base communities would numerically overshadow regular churches. Apart from Brazil, however, the base community movement remains small. What have proved to be more dynamic, at least numerically, have been Protestant evangelical communities and, increasingly, Pentecostal groups. Nonetheless, the base communities are a significant feature of the ecclesial landscape of South and Central America, a fact which has prompted many liberationists to shift their hope for the future away from Marxist revolutionary theory to the leavening activity of the base communities. As Paul Sigmund puts it in an analysis of recent developments within liberation theology, there is a change "from an infatuation with socialist revolution to a recognition that the poor are not going to be liberated by cataclysmic political transformations, but by organizational and personal activities in Base Communities."[12]

This emphasis on the base communities has been influential in garnering support from European and North American Christians who were sympathetic to the plight of the poor in the southern hemisphere, but who worried about the heavy Marxist emphasis in liberationism. The influential theologian Jurgen Moltmann once criticized liberation theology for its captivity to Marxist theory in a famous letter. The rise of the base communities prompted a change of heart.

The theologians of liberation have sunk their roots in the base communities, which are a promising sign of reform of the church and society, and which inject life in a way that has something of the miraculous about it in a rather apathetic, centralistic church. And that is where the theology of liberation now has its organizational relationship. Different from Marxism? I should say so! Today, then, I can state that liberation theology is a solid, sound theology, and I drop, part and parcel, the complexities I formulated in that letter of mine. I know that many objections exist against the theology of liberation, especially with regard to the popular church. However, I believe that the experience being lived under the name of the popular church is a new experience of the Holy Spirit – new experience of Pentecost.[13]

THE FATE OF THE
LIBERATIONIST MOVEMENT

By the early 1990s, this new Pentecost, which began in 1970, was beginning to lose much of its luster. As dependency theory vanished, the Berlin Wall was picked apart, and the Soviet Union unraveled like a cheap sweater, liberationists and their sympathizers began to question the future of the movement. When I set out to write this book on new movements in science, religion and spirituality, the two movements which most people thought did not belong in such a book were fundamentalism and liberation theology. Fundamentalism, however, has since proven its resiliency and its ability to adapt to different countries and different faith groups. Instead of dying away it seems to have grown more pervasive, like a United States corporation that has suddenly gone multinational. In comparison, liberation theology has appeared to have died down, having lost a focus for its attacks. In the 1980s Harvey Cox fingered fundamentalism and liberation theology as the two movements to watch in the shift from a modern to a postmodern world; in the 1990s his candidates were fundamentalism and Pentecostalism.

I, for one, do not think that liberation theology will disappear for some time. In the first place, the economic plight of most people in the Two-Thirds World has not improved. Slums bloom in Brazil, which not long ago was touted as the next great economic powerhouse. Even in Canada where I live and work, the gap between poor and rich has widened and the middle class is slowly dwindling. Moreover, liberation theology has spun off its liberationist themes into Black theology, feminist theology, gay theology, Hispanic theology, and Native American theology.

Often these related liberationist movements have furious disagreements over the locus of oppression and the consequent goal of liberation. They are bound together, though, by a common set of presuppositions that give musical breadth to what was a one-note symphony. Robert McAfee Brown provides a chart that highlights the differences between traditional western theology and this new theological enterprise:

TRADITIONAL THEOLOGY	LIBERATION THEOLOGY
1. begins with the world of modernity and remains thought-oriented	1. begins with the world of oppression and becomes action-oriented
2. responds to the nonbeliever whose faith is threatened by modernity	2. responds to the nonperson whose faith is threatened by forces of destruction
3. is developed "from above" – from the position of the privileged, the affluent, the bourgeois	3. is developed "from below" – from the "underside of history" the position of the oppressed, the manipulated, the exploited
4. largely written by "those with white hands," the "winners"	4. only beginning to be written, must be articulated by those with dark-skinned, gnarled hand, the "losers"
5. focuses attention on the "religious" world that needs to be reinforced	5. focuses attention on a political world that needs to be replaced
6. linked to Western culture, the white race, the male sex, the bourgeois class	6. linked to the "wretched of the earth," the marginated races, despised cultures and sex, the exploited classes
7. affirms the achievements of culture – individualism, rationalism, capitalism, the bourgeois spirit	7. insists that the "achievements" of culture have been used to exploit the poor
8. wants to work gradually, reforming existing structures by "supervision"[14]	8. demands to work rapidly through liberation from existing structures by "subversion"

Although, I neither agree completely with Brown's analysis of the differences between the liberationist perspective and that of traditional Western theology, nor with all aspects of liberation theology, I must admit to a personal sympathy with liberationism, particularly as articulated in Latin America. I have seen too much mind-numbing poverty in the streets of La Paz and the Altiplano countryside not to support any theology which attempts to rectify the economic inequalities operative in that, or any other country. Moreover, any nonbiased reading of the Bible reveals God's concern for economic equality. As Gandhi is said to have stated, the Bible makes it clear that there is enough in this world for everyone's need but not enough for everyone's greed.

A poll taken in Brazil in the early 1960s revealed that two-thirds of university students in Rio de Janeiro considered themselves to be unbelievers who saw the church as an instrument of oppression. A poll taken some 15 years later in 1978 showed a striking change, "with three-fourths of students declaring themselves to be believers and favorable to the church."[15] Clearly, this theological movement has struck a responsive chord among many!

AN ASSESSMENT

In spite of heartfelt sympathies, though, liberation theology fails on two important levels. The first is due to Marx's historical determinism, based on his reading of Hegel's writings. This surfaces in a recent book by Alistair Kee in which he criticizes the liberationists for not being Marxist enough! For Kee, this is a source of great disappointment as he states, "Liberation theology has been criticized for being too Marxist: in reality it is not Marxist enough. Its allegiance to Marx has been regarded as the basis of its success: in fact its resistance to Marx is the cause of its failure."[16] The problem, according to Kee, is that "according to historical materialism it is not possible to leap from a feudal society to a communist society."[17] Kee's criticism is true on one level – capitalism has not taken hold in many sectors of Latin America, and therefore, the liberationists are attacking a nonexistent economic system.

The failure, though, is mainly a failure of Marx's dialectic and the fact that the liberationists have followed it too slavishly; they have bought into Marx's problematic dualism.

Hegel, himself, I am convinced, was not nearly as wooden as Marx made him out to be. He did not believe, as Marx did, that the spirit worked in a blind inexorable fashion. Marx's world spirit, based on a dualism of thesis versus antithesis (the proletariat versus the bourgeoisie), hobbles the Spirit. Unlike Jesus' comment that the wind blows where it will and you hear the sound of it but do not know from where it comes or to where it goes (a comment which he applies to the activity of the spirit), Marx's spirit is tame and predictable: *this* lead to *that*, which is followed by *this*, in a logical, linear fashion. Marx was proud that his theory was scientifically grounded, but in an age of quantum physics is his theory as scientifically grounded as it was in the Newtonian era? This is a good question and one reason why I welcome the liberationists' move toward a more dynamic, pluralistic view of the Spirit, in tune with the Bible rather than with Marx.

The second failure in the liberationist vision is the type of society that they put forward as their utopia. Westerners living in suburbia can be forgiven for not finding the liberationists' visions inspiring. While adequate food and shelter is important, it is really just a precondition for something deeper, a hunger of the spirit which cannot be fed by bread alone.

I can still remember my first trip back to North America after five years in Bolivia. I visited a friend whose father lived near Washington, D.C., awaiting posting with the diplomatic corps after a period of debriefing. My friend took me to the local shopping mall and we went into an electronics store. From a country where there was no television, I was suddenly confronted with a wall of televisions, all turned on, and all in color. For years afterward, I would dream of that wall of televisions and in my dreams I could even visualize the color of the program which was playing. I wanted that life. I became a typical consumer and almost lost my spirit in the quest for the good life as defined by material wealth.

The world will not be saved when everyone is able to buy the latest color TV, with remote control and cable access to 500 channels. To paraphrase the Russian novelist Aleksandr Solzhenitsyn, the goal of human existence is not the accumulation of material goods, but rather spiritual growth. The liberationists turn a social precondition into a final goal and thereby reduce the Spirit in the process of trying to save it. It is not liberationism, then, which can provide for Western society the necessary social vision, but an offshoot of liberationism – religious feminism and the larger feminist movement.

9

RELIGIOUS FEMINISM

On the evening of April 23, 1976, several hundred women came together to participate in the first national all-woman conference on women's spirituality. The keynote speeches and opening rituals were held in a church in the heart of Boston. After listening attentively to two addresses on the theme of "Womanpower: Energy Re-Sourcement," the audience became very active. In tones ranging from whispers to shouts, they chanted, "The Goddess is Alive – Magic is Afoot." The women evoked the Goddess with dancing, stamping, clapping and yelling. They stood on pews and danced bare-breasted on the pulpit and amid the hymnbooks. Had any sedate, white-haired clergyman been present, I am sure he would have felt the Apocalypse had arrived.[1]

The religious feminist movement has been one of the most profound and important movements in the West and, increasingly, throughout the world. It is a movement for which I have profound respect and, at the same time, deep ambivalence. It is a movement that indirectly has hurt my career hopes. At the same time, it is a movement whose vision has been a source of healing.

I am not alone in my assessment of the importance of religious feminism. Alistair Kee in his book on liberation theology touts liberationism as the most important theological movement to have arisen in recent years, second only, he hastens to add, to feminist theology.[2] Richard Tarnas, in a fascinating overview of Western civilization, concludes that the crisis in which we now find ourselves will only be resolved as we take feminist criticism seriously and repent of the sexism of the past 2,000 years and more.[3]

Religious feminism is the third and last religiously inspired movement with a comprehensive social vision to be examined in this book. Its placement at the end is not accidental. Historically speaking, it would be a toss-up as to whether feminism or fundamentalism had prior claim. However, the ordering of these three movements was not historically inspired. I have admiration for many elements of the fundamentalist social vision, particularly its rejection of the sacred-secular dualistic split that has so crippled religious faith. I admire even more liberation theology's vision of an equitable political and economic order, where children do not die of starvation and where women and men are not worn out simply trying to survive. However, neither of these two movements has the breadth and the depth that can provide a social vision for the new religious insights and spiritual awareness coming to birth in the West. Only feminism, particularly religious feminism, has the breadth of vision necessary to incarnate the spirituality of the third millennium.

I realize that feminism, even religious feminism, is not a monolithic entity. Some aspects of religious feminism are silly, and some are dangerous and even hateful. The vision, however, common to most forms of religious feminism is larger than individual feminists or feminist groups. In the end, the closest parallel I can find to the feminist vision is one which the prophet Isaiah gave voice to many centuries ago when he foresaw a time when relationships between the human creature and the animal world, the human creature and the environment, the human creature and God would be healed (Isaiah 11:6–9). This is the only vision I am aware of which is worth spending one's life and efforts to achieve. At their best, this is the social vision which religious feminists advocate.

As with liberationists, feminists have had to rewrite history, or as they prefer to call it *herstory*, since it is a truism that history is written by the victors. In the case of the liberationists a long-lost historical figure, Bartholomew De Las Casas, a monk who worked with the indigenous peoples, has provided much encouragement for the present. For religious feminists, particularly Christians, an intriguing 12th-century German abbess named Hildegard of Bingen has functioned as a source of inspiration.

Known for her prophetic visions, Hildegard counted popes and emperors among her acquaintances. Her career was a rich one. Despite illnesses, she wrote extensively on a diverse number of subjects. Today St. Hildegard has been resurrected. Her writings have been reproduced; her music has come back in fashion.

To Hildegard could be added Spanish nun St. Teresa of Avila in the 1500s and French nun St. Theresa of Lisieux in the 1880s, serving as a helpful reminder that history cannot be divided off into nice, neat sections as historians love to do.

THE RISE OF RELIGIOUS FEMINISM

The warning having been sounded, it is fair to claim that the modern feminist movement in North America began with the work of Elizabeth Cady Stanton. Stanton was a veteran of the 19th-century antislavery campaign as a result of her marriage to the noted abolitionist Henry Stanton. More to the point, she was an ardent feminist. Born in Johnstown, New York in 1815, she graduated from Emma Willard's Female Seminary in Troy, New York, in 1832. In the 1840s, Stanton campaigned successfully for the passage of the Married Women's Property Act in New York State. In 1848, she organized the first women's rights convention, convened at Seneca Falls, and helped draft its "Convention of Sentiments" which called for women's suffrage, divorce reform, and access to professions. An indefatigable worker, in her 80s she began work on *The Woman's Bible*, a commentary on the Bible intended to correct the prevalent anti-female interpretation from which the Bible suffered. Although controversial, *The Woman's Bible* became a bestseller, going through seven printings in the first six months as well as translations into several different languages. Unfortunately, at the end of her life, Stanton became convinced that there was little in the Bible and thus in Christianity which was of value for women.

Stanton's conclusions meant that the next phase of the feminist movement was not concerned with spirituality, but with political reform and with the issue of suffrage. This narrowing of the feminist vision was to have negative results for women in general and for society as a whole. According to Anne Carr, "By placing their hope in the almost miraculous changes they thought would come about through the vote, nineteenth-century women lost their wider vision of change in all aspects of women's life. *The vote for women was won in 1920, and the women's movement died.*" [emphasis mine][4]

It was not until the 1950s and, especially, the 1960s, that the feminist movement would be reborn. The decade began with the striking of the Presidential Committee on the Status of Women. Just as important for religious feminism, however, was the publication in that same year of an intriguing article by Valerie Saiving Goldstein. In "The Human Situation: a Feminine View," Goldstein argued for a reconsideration of the categories of sin and redemption prevalent within the Judeo-Christian tradition. From the time of the apostle Paul, the chief sin was thought to be the sin of pride/hubris in which the human creature overreached himself and attempted to be like God. Goldstein claimed that while this may have been true of the male gender it was not true of the female gender.

Governed by the Sisyphean myth rather than the male Promethean one, females were afflicted by the sin of sloth rather than pride. In the myth of Sisyphus, the lot of the individual in life is compared to Sisyphus, who must roll a large stone up a hill. Just as Sisyphus reaches the top, the rock rolls down again. Over and over again, the scene is played out – not a bad analogy for housework! In contrast, Prometheus dares to steal fire from the gods, which he then gives to humanity. Instead of believing in themselves and attempting great things like Prometheus, women were hesitant and timid. They did not need to hear sermons and read books on the sin of pride, they needed to hear sermons and read books on sloth. Emphasis on pride only served to further enslave women and retard their growth.

Goldstein's article served not only as an example of how to reinterpret traditional religious dogma, but also as a call to action. It was joined, in 1965, by an important article by Mary Daly, at that time an active Roman Catholic scholar. In 1968, Daly's article was expanded into a book, *The Church and the Second Sex*. An impassioned writer, Daly built on the progress made in Vatican II, but was critical that it had not gone any further in its reforms. While at this time she still counted herself within the fold of the

Roman Catholic church, there were indications that Daly viewed not only the church but Christianity itself as irredeemably sexist. By 1973, Daly had concluded that she could no longer function within the confines of the church and had made her famous pronouncement, "As long as we believe God is male, then the male thinks he is God."

In tandem with what was happening within religious feminism were changes and proposals made by feminists addressing social and political issues. During the 1960s and 1970s, the focus of such activists was directed toward equality in the work force, specifically equal access to jobs, equal opportunity for advancement, and equal pay for equal work. The work of the religious feminists, however, along with the influence of the theology of liberation, pushed the feminist movement in the 1980s and 1990s in the direction of a "more profound critique of the cultural, domestic, economic, and political systems that have hindered and devalued women."[5] This re-valuation of the systems which have oppressed women has led to a plural-ism within the feminist movement.

Originally characterized by white, middle- to upper-middle-class lead-ership, the main tension in the earlier decades was between feminists who wished to reform the church from the inside and those who had no use for religious institutions or, in the secular 1960s, religion itself. By the 1980s and 1990s, these tensions led writers such as Mary to move so far beyond the confines of the church and Christian faith that a controversy developed between those who felt that Christianity could be rescued and those who felt that it could not and that in its place neo-pagan religions needed to be substituted. Daly added to this a rejection of male-female relations, affirm-ing lesbianism as the best and psychologically safest lifestyle for a woman.

According to Daly, men could be characterized as "plug-uglies" who, although they give the illusion that they are giving something, like the elec-trical male plug are always taking. As Daly puts it, men are "drainers of energy whose plugged-in fittings close women's circuits, sapping the flow of gynergetic currents so that these cannot circulate within/among women."[6] Daly equates the Marquis de Sade's "sadistic pornography" with the "sado-masochistic theology of Karl Barth." From her lesbian, neo-pagan perspective, they are "essentially the same."[7]

Daly is on the extreme of the religious feminist continuum. As she says of herself in an imaginary inquisition between a representative of the Roman Catholic church and her two cats, who respond to the criticism that Daly has become heretical, "you mean *Her-etical*, or *Her-Ethical. Doctor Daly has become too Weird to be classified merely as 'heretical'.*" [emphasis mine][8] Another religious feminist who has joined Daly in rejecting Christianity is more representative of this wing of the feminist movement. While the one-time Anglican, Daphne Hampson, has left the church behind, she does not share Daly's distaste for males.

It should be pointed out once again that those feminists who say that, on account of male symbolism, Christianity is impossible for them, are not thereby necessarily "anti-men." It is a retort which is sometimes given, but quite beside the point. For we are not here, in speaking Christologically, conceiving of Jesus as simply one human being among others who lived in the past. Christianity gives a male human being a status which is given to no woman... Indeed it may be (as in my case) because one deeply cares that there should be good and equal relations between men and women that one is adamant that no one human being can be given the kind of status which Christians give to Christ. Such a religion as Christianity is a symbolic distortion of the relationships which I would have... I am therefore not a Christian not least for the sake of the kind of human relationships which I want to see flourish.[9]

The religious feminist movement has not abandoned Christianity altogether, though. Christian feminist such as Canadian Pamela Dickey-Young, who teaches at Queen's University, and represents the opposite end of the feminist continuum from that of Mary Daly, speaks for many when she claims that Christian feminists "must be willing to enter fully into the meaning of the Christian tradition without compromising their feminist ideals. Yet they accept neither Christian tradition nor feminist ideals uncritically."[10] In this group belongs Elizabeth Schüssler Fiorenza, a renowned biblical and feminist scholar identified with the Roman Catholic church. A third group between Hampson and Dickey-Young is best represented by the prolific author and scholar, Rosemary Radford Ruether. As opposed to Hampson who feels that Christianity is not reformable and Dickey-Young who claims that it is, Ruether holds that Christianity is reformable but that extrabiblical sources taken from neo-pagan traditions must be added to the Christian tradition in order to make it palatable for women.

WOMANISM

From my viewpoint, as one who in a round-table discussion was criticized for raising the issue of class as more important than the issue of gender, one of the most significant aspects of the feminist movement recently has been the rise of Black feminism, or womanism, as many African-American feminists like to call it, in order to differentiate their perspective from that of the feminist movement which they feel has been representative only of upper-middle-class, white women. For many African-American womanists, white feminists were dealing with *fulfillment* issues while they have had to deal with *survival* issues. As theologian Jacquelyn Grant thunders out:

...White women have defined the [feminist] movement and presumed to do so not only for themselves but also for non-White women. They have misnamed themselves by calling themselves feminists when in fact they are White feminists... To misname themselves as "feminists" who appeal to "women's experience" is to do what oppressors always do; it is to define the rules and then solicit others to play the game. It is to presume a commonality with oppressed women that oppressed women themselves do not share.[11]

One current issue in the debate between North American White feminists and others within the feminist movement concerns the practice of female circumcision in non-Western cultures. While the majority of feminists, of whatever color, condemn the practice, there are a few lonely voices raising questions of cultural domination by the West under the guise of promoting feminist values. Recalling the movie *Hawaii*, based on James Michener's book of the same title, in which white male missionaries tried to get the indigenous people of Hawaii to dress appropriately and practice monogamy, it is sometimes difficult not to feel that feminists are practicing a contemporary form of cultural imperialism. Although one must hasten to add that the two things compared here – the wearing of appropriate clothing (an inconvenience) and the practicing of female circumcision (a life-threatening practice) – are not equal in magnitude.

A RENEWED EMPHASIS ON SPIRITUALITY

One of the most interesting aspects of contemporary feminism is the return to spirituality, which was so much part of the early feminist agenda. The secularism of the 1960s and 1970s has given way to a renewed emphasis on spirituality, most of it non-Christian. In particular, neo-pagan faiths such as Wicca have served as a vehicle for the expression of women's spirituality. The rebirth of witchcraft was not originally tied in with feminism; it was spearheaded by Gerald Gardner, a British civil servant, who claimed that in 1939 he had been initiated into a coven of witches. Gardner's Wiccan movement, however, has proved attractive to women because of its emphasis on worship of the Goddess. Some scholars depict this spiritual renaissance as the last gasp of the feminist movement; other more sympathetic observers call it the "third wave" of feminism. The first wave was the initial work for women's political rights, particularly the right to vote. The second wave was the agitation for equality of pay, advancement, and employment. The third wave is the birth of feminine spirituality freed from the constraints of a Christianity deemed to be misogynist and sexist.

This third wave is not just a minor aspect of the feminist movement as one might think at first. In a class I teach for Mount Allison University in Sackville, New Brunswick, one of the assignments deals with a comparison between the fertility religion of Baalism, which the Israelites encountered upon entering the Promised Land, and the Jewish faith. Almost 90 percent of my female students express a preference for Baalism over Judaism simply on the basis that Baalism has a female Goddess while Judaism and Christianity, to their thinking at least, worship a male God. Because the class is a first-year introductory course and most students admit they have never studied religion previously, I can only conclude that the third wave of feminism has influenced their thinking almost subconsciously.

THE INFLUENCE OF LIBERATIONISM

Since religious feminism, especially as it developed in the West, was so heavily influenced by the theology of liberation, it is not surprising to find that many of the key tenets of religious feminism are similar to liberationism. To begin, religious feminists accept the insight that one's context determines one's theological understanding. As a consequence feminists are critical of metastories which they feel serve to oppress rather than to liberate. Metastories are stories that attempt to give a unified, systematic overview. In opposition, women have directed their attention to individual stories of women that have been ignored by these metastories.

During one of my doctoral courses at McGill University in Montreal, the professor was seeking our opinions on the life and thought of St. Augustine. One of the women doctoral students prefaced her remarks by saying, "As a woman, I think..." At this the professor, an older white male, burst in caustically: "I don't care what you think as a woman, I want to know what you know about Augustine." According to religious feminists, that professor was operating out of a metaperspective which does not take seriously the context of oppression under which women have labored. Systematized in the form of universal reason and history, the only way women have been able to expose the prejudice of the past is by focusing on their individual experiences and those of other women.

As they have focused on their context, almost all religious feminists draw the conclusion that not only the Christian church but Christian theology as a whole is patriarchal, written by men and assuming maleness as the norm. Moreover, when it has dealt with women, this theological tradition has caricatured women's experience and been harmful to the full humanization of women.

An example of the patriarchal nature of the Christian church comes easily to mind. I was working as the student minister at a church in Toronto, along with my wife who served as the Christian Education director. Up to that time, the church had only had male deacons, the governing body in a Baptist church. However, at the annual meeting a group of people indicated their desire to elect women to the deacons' board. Since the senior minister was against this decision, my wife and I had to keep a low profile, although we quietly made our position known.

I can still recall one stormy deacons' board meeting which my wife, although she worked full time and I only part time, was not allowed to attend. One deacon rhetorically asked, referring to the passage in 1 Timothy about a church leader being the husband of one wife, "How can a woman be the husband of a wife?" The chair of the board at that time was a single man, but apparently this passage, even when interpreted in a literal, wooden fashion did not apply to him because he was male. At that point another deacon, a kind and caring man in other worlds, burst out caustically, "If women become deacons then all we'll end up doing in our meetings is debating the color of the curtains in the church parlor!" His comment cut two ways. In the first instance, he assumed that women were only interested in stereotypical roles. In the second, he assumed that the color of the curtains was less important than other issues, even though much of our meetings were spent on minor details. Apparently making sure the church lawn mower was in good repair was a legitimate concern while the color of the parlor curtains was not!

The context of exclusion and oppression in various forms is the starting point for Christian feminists and, since the same holds true for other religious faiths, for religious feminism as a whole. Religious feminists, however, were never wedded to orthopraxy (correct action) as opposed to orthodoxy (correct doctrine) as were the liberationists. In part, this was because feminism needed to construct a new orthodoxy since the previous one was shot through with sexism. Nonetheless, the stress on experience was a predominant one. Elizabeth Schüssler Fiorenza, in a short autobiographical sketch, notes that when she began her academic career she "assumed that the experience of women and the praxis [action] of church and ministry should be primary for articulating ecclesiology and spirituality." As Fiorenza attempted to rethink theological constructs from the marginal location of a "lay woman" in the Roman Catholic church, she became aware of her "marginalization". In the late 1960s and early 1970s, the role of women in the academy was so uncertain that Fiorenza's academic supervisor refused to obtain a scholarship for her in spite of her excellent academic achievements because "*he did not want to waste the opportunity on a student who as a woman had no future in the academy.*" [emphasis mine][12] Today the shoe is often on the other foot, as white males find it increasingly difficult to land teaching positions in most arts subjects.

The preferential option for the poor, another vital plank in the liberationist's platform, becomes the preferential option for women in religious feminism. The interpretation of this preferential option, however, is not simply a concern for women and their well-being. It is, rather, the epistemological option of women; the claim that, because of their marginalization within religious structures and religious beliefs, women are better able to see the institutional and ideological changes that have to be made. White feminists already knew from experience that women could be cared for in a way that ensured their physical needs, but starved their spiritual and emotional needs. They moved quickly past the liberationists' earlier concern for the poor as being patriarchal to the more radical *epistemological* option for the women whereby women not only deserve special treatment but possess special, superior insight. At times this has marginalized men, as women have had to make room for their views to be heard. Consequently, women have created female-only groups known as women churches in order to further the process of unmasking patriarchal traps within Western society and, more importantly, to reimage society freed from gender bias.

There are many ways to engage in this process of reimaging. Chinnici provides one such method. Similar to that used by the liberationists, the process forms an interpretative circle where new insights affect one's understanding and thus begin the process of reimaging again. Chinnici's advice is to 1) select an experience upon which you would like to reflect; 2) reflect on this experience from principles of feminist theology; 3) attempt to name this experience with a *traditional* word; 4) study the tradition of the word you have chosen; 5) begin the process of reimaging.[13]

REIMAGING THE BIBLE

One of the tasks which religious feminists in the West have had to engage in, almost from the very start, is reimaging the Bible. Stanton concluded that this was an impossibility. Contemporary authors such as Daly and Hampson substituted various forms of neo-paganism for Christianity and Judaism. Others, however, felt that the Bible could be successfully reimaged. For feminists this was a far more complex process than for liberationists. Most of the time liberation theologians simply had to point to the biblical text and ask why we North Americans and Europeans did not read the literal text and put it into practice. Although some texts such as Paul's Letter to the Galatians (in which he claims that because Christians are all one in Christ there is neither male nor female) could be treated like this, other texts could not. There are texts in both the New Testament and the Hebrew Bible (Old Testament) which definitely assign a secondary place to women.

The Temple system itself illustrates this secondary status of women. At the core of the Temple was the Holy of Holies into which only the High Priest could enter. Outside this was the courtyard of the priests, open only to the priestly class. Next was the courtyard of the Jewish men and only then, was there assigned a place for women. Moreover, from our contemporary perspective it means nothing to us that in the biblical story Eve ate of the fruit of the tree of knowledge of good and evil before Adam did; in traditional Jewish exegesis, with its emphasis on the importance of the first-born, this has crucial implications as it suggests that sin began with Eve – the first woman.

In the New Testament, in spite of Jesus' example of inclusivity toward women, there are some troubling passages traditionally assigned to the apostle Paul which teach that women should not speak in church and should defer to male spiritual authority. As a missionary kid, I always found this confusing. Women missionaries were some of the great heroes of my childhood. In Bolivia they could preach, teach, and do almost anything. In Canada they were regulated to the pew or to missionary movements but were almost completely excluded from ordained ministry. Paul's comments provided the biblical justification for this, along with the fact that Jesus was a male and the 12 apostles were all males.

The reimaging of the Bible advanced along several fronts. Technical studies as well as larger theological studies on the meaning behind the biblical words were brought into service in order to retranslate the Bible in a way that would be freeing for women. Increasingly, as they studied and reflected on the biblical text many religious feminists found they could not reimage the Bible satisfactorily. Rosemary Radford Ruether led the way, insisting that,

Feminist theology cannot be done from the existing base of the Christian Bible. The Old and New Testaments have been shaped in their formation, their transmission, and, finally their canonization to sacralize patriarchy. They may preserve, between the lines, memories of women's experience. But in their present form and intention they are designed to erase women's existence as subjects and to mention women only as objects of male definition.[14]

Along with adding extrabiblical texts to the feminist mix in order to compensate for the overwhelming patriarchal bias in the Bible, Ruether also provides liturgical services for exorcism of texts, which cannot be reinterpreted in a healthy, inclusive fashion.

EXORCISM OF PATRIARCHICAL TEXTS

A small table with a bell, a candle, and the Bible are assembled in the center of the group. A series of texts with clearly oppressive intentions are read. After each reading, the bell is rung as the reader raises up the book. The community cries out in unison, "Out, demons, out!"

Suggested texts in need of exorcism:

Exodus 19:1,7–9,14–15 (shunning of women during giving of the Law at Sinai)
Judges 19 (rape, torture, and dismemberment of the concubine)
Leviticus 12:1–5 (uncleanliness of women after childbirth)

Ephesians 5:21–31 (male headship over women compared to the relation of Christ and the Church)
1 Timothy 2:11–15 (women told to keep silence in church and to be saved by bearing children because they are second in creation and first in sin)
1 Peter 2:18–20 (slaves exhorted to accept unjust suffering from their masters as a way of sharing in Christ's crucifixion)

At the end of the exorcism, someone says, "These texts and all oppressive texts have lost their power over our lives. We no longer need to apologize for them or try to interpret them as words of truth, but we cast out their oppressive message as expressions of evil and justifications of evil.[15]

Another problem in the area of biblical exegesis which feminists had to deal with, which was not an issue for the liberationists, was the gender of God. If the first few chapters of Genesis are taken as a guide, God is beyond gender; the first creation story (Genesis 1:27) specifies that both men and women were created in the divine image. However, the English language in particular has been problematic, as gender-specific language has been used for God. At one time, the male pronouns were considered generic; *man* was not just the male gender but the prototypical human being. In a similar fashion, the use of *He* when applied to God was not meant to designate that God was male but was used generically. Unfortunately, many Christians do not treat the pronoun as a generic reference but interpret it literally. Ironically, many feminists also interpret such language in this same way. In this they may agree – the biblical God is a male God – although they disagree on whether or not this makes the Christian God a suitable deity for females to worship.

I disagree that the male pronoun is anything but generic; even Jesus' reference to God as *abba*, the Aramaic endearment for "father," is a generic reference to the fact that God relates to us in a loving, parental fashion. Nonetheless, I became sensitized to the problems of language at my church in Kingston, Ontario, as well as with my own two daughters who have only very recently, in their late teens, come to realize that God is not a male. Somehow the use of the words "father," "lord," "king" spoke louder to them than any disclaimers.

I also remember graphically an incident that took place at an open session of Sunday school in a former church. The Sunday school superintendent was speaking on God as a loving, caring father. Everything would have been okay except this person wanted to press her point home, so she singled out a student whose father was Canada's most famous musician and concluded, "God is just like your father, Jimmy."[16] At that, Jimmy, a 12-year-old boy, burst out, "My father isn't loving and kind. He drinks and he beats my mother." Everyone in the room received a very pointed lesson on the pitfalls of gender-specific language when applied to God! Elizabeth Johnson sums up the situation and the problem, noting that,

To even the casual observer it is obvious that the Christian community ordinarily speaks about God on the model of the ruling male human being. Both the images used and the concepts accompanying them reflect the experience of men in charge within a patriachal system. The difficulty does not lie in the fact that male metaphors are used, for men too are made in the image of God and may suitably serve as finite beginning points for reference to God. Rather, the problem consists in the fact that these male terms are used exclusively, literally, and patriarchically.[17]

WOMEN CHURCHES

The emphasis on experience, the awareness of systemic sexism within church and synagogue, the belief that much of the Bible is patriarchal, and the problem of gender-specific language propelled many religious feminists in two directions. The first was simply to leave the Jewish or Christian faiths for a neo-pagan alternative or for atheism. The second was to create a parallel structure to the religious institutions of the day. Often participants in these feminist churches would also belong to traditional church structures, leading to a sense of religious schizophrenia. Typical of such feminists is a woman, Anglican by birth, active on her parish council and yet active in extrachurch institutions. She writes:

...in December I will be actively involved in two very different spiritual events. Once again, I will be a narrator for several performances of an absolutely traditional retelling of the Christmas story, with music and mime and narration that is taken directly from the gospels, King James Version... And also, once again, I'm one of a small group of women who create our semi-annual solstice gathering, WomanSong, which brings together between 200 and 300 women of all traditions and no tradition; in this event we honor the goddess and the feminine expression of spirituality, and we celebrate with music, ritual, meditation, reflection and feasting... I attended a workshop led by Jean Shinoda Bolen a couple of months ago; she refers to herself as an Episcopagan and I find that a very useful term.[18]

These women churches are patterned after the base communities of the liberationists. Typically, they have elicited a similar condemnation from traditionalists who feel that such a movement undermines the institutional church. With the liberationists, it has been the Roman Catholic hierarchy which has advanced this criticism. In the case of the Women-Church movement, two conservative Protestant theologians warn that feminist theology, "threatens a new schism within the body of Christ by its support and encouragement of the Women-Church movement."[19]

In these women churches, participants have felt free to explore liturgical practices that would be frowned on within the established religious organizations. More importantly, they have felt free to worship the Goddess of the Earth rather than the God of the Sky, as many feminists typify the biblical deity. In this, they have been encouraged by archaeological findings claiming that worship of a mother Goddess preceded worship of a father God. These findings are not nearly as unequivocal as many feminists treat them, as indicated in chapter three on Gaia. Nonetheless, it is true that polytheistic faiths are ancient and that monotheism was a later, albeit very old, development in its own right. Most of the time the polytheistic faiths were tied to the fertility cycle of the Earth and thus by a logical inference to the fertility of women.

The various forms of Baalism prevalent in the Promised Land prior to the Israelite conquest, and even long after, are representative of such faiths. In Baalism, there were at least two Gods – a male and a female deity. The copulation of the Gods was thought to result in the fertility of the Earth. As we moderns seed the clouds in order to produce rain, so the worshippers of this ancient fertility religion felt that by sympathetic magic the Gods could be encouraged to copulate, thus ensuring an abundant harvest. This explains the growth of temple prostitution which the Israelite prophets railed against.

A chart comparing the salient features of both types of religious faiths illustrates the vast differences between the two, as well as some of the features which might be considered attractive to contemporary feminists:

Canaanite Religion (Baalism)	Early Judaism
1. nature mysticism	1. history mysticism
2. time is cyclical-repetitive	2. time is progressive and linear
3. emphasis on immanence	3. emphasis on transcendence
4. dualistic	4. monistic
5. female goddess	5. no female goddess
6. sacred sexuality	6. sacred sociality
7. liturgical emphasis	7. ethical emphasis
8. control of gods by imitative magic	8. covenantal relationship
9. utilitarian (if it works it's right)	9. non-utilitarian
10. polytheistic	10. monotheistic

The stark contrast between these two religions has prompted the noted scholar Elizabeth Achtemeier to issue a stern warning about the worship of the Goddess. She exclaims:

No religion in the world is as old as this immanentist identification of God with creation. It forms the basis of every nonbiblical religion, except Islam; and if the church uses language that obscures God's holy otherness from creation, it opens the door to corruption of the biblical faith in that transcendent God, who works in creation only by His Word and the Spirit. Worshippers of a Mother Goddess ulti-mately worship the creation and themselves rather than the Creator.[20]

Worship of the Goddess is an important problem area in religious femi-nism, though not for the reasons which Achtemeier proposes. In the Chris-tian tradition, God is conceived of as further away than the furthest star but also closer than the beat of your own heart. God is both transcendent and immanent, and a Christianity which emphasizes the transcendence of God to the exclusion of God's immanence is as distorted a view of the divine as one that completely identifies God with the creation. The problem, then, is not with the immanentist aspects of Goddess worship, but with the frag-mentation of the Oneness of the divine into two Gods, one male and one female.

THE GOD-GODDESS DUALISM

This dualism is no healthier than that found in Protestant fundamentalism and often leads to the same exclusivistic attitude. I have met many femi-nists, particularly lesbian feminists, who were just as rigid, unloving, and judgmental as any fundamentalist I have known. Noted feminist Elizabeth Schüssler Fiorenza warns that while feminists "criticized binary opposi-tions and asymmetric dualism, [they] nevertheless tended to sustain such dualism by conceptualizing patriarchy in terms of gender antagonism and male-female oppression rather than in terms of the complex interstructuring of sexism, racism, class-exploitation and colonialism in women's lives."[21] Another feminist scholar, Rosemary Radford Ruether, cautions that anger against men may well be an important stage in the consciousness-raising of the typical feminist, but it must end there, as "one needs to recognize one's own fallibility, one's own capacity not only to be victimized, but also to be victimizer, and, in the mature self-esteem, also be able to affirm the hu-manity of males behind the masks of patriarchy.[22]

I suppose it could be argued that this new phase in the feminist move-ment does not advocate two Gods, one male and one female, but a plurality of gods in which the Goddess is one among many. However, a true pluralism must always be bound together by a unity of some sort or the pluralism cannot continue to exist. In the panoply of the gods of the various polytheistic religions, always one or two gods surface as more important than the rest. The problem is that usually two gods vie for pride of place and we are back to the dualism, which has been such a problem.

Another argument might be that the unity rests not in the deities, but in the feminist ideology. However, once this perspective is adopted, the feminist movement suddenly loses the diversity that has characterized it to date, a loss that I personally would lament. An illustration might help as to why this would happen. Within the Roman Catholic church there is an incredible array of theological diversity. To the uninformed, it often seems like the Roman church is a monolithic theological institution but nothing could be further from the truth theologically. The uniformity in the Roman Church rests in the structure, not in the theology. In comparison, conservative Protestant churches have a diversity of organizational structures, but because of that insist on a theological uniformity, which for a free thinker is extremely repugnant. In many conservative Protestant churches, the minister and congregation have to subscribe to a statement of faith; they can practice all sorts of forms of worship but they must be doctrinally united, no deviance is permitted.

Much more fruitful than the polytheistic option of two Gods is the approach taken by most Christian feminists attempting to look beyond the masculine images in which God has been clothed to the true essence of the divine being. The fact that the book of Genesis explicitly states that male *and* female are created in God's image has been a great help in this regard, as have various passages sprinkled throughout the Bible where God is depicted in feminine terms (cf. Hosea 11:1–9).

A second problem which has afflicted the feminist movement at large as well as religious feminism has been the loss of its social vision. This loss can also be attributed in large part to the dualism creeping back into the movement. If men are the problem, then the solution is not seen to rest in a renewed relationship between the genders, but in women claiming their rightful power and putting men in their proper place.

An equally important reason for this loss of vision, however, has been the siren call of our consumer society. Some women have tasted the good life and in doing so have forgotten that the vision of feminism is a vision for the whole. This is clearly shown in the emphasis one finds upon women's success stories. Originally, the sharing of such stories within feminist circles was not to elevate the protagonist, who managed to gain success through grace rather than effort, but the group. As more and more women managed to break through the glass ceilings of professional advancement, this emphasis paled in the face of an emphasis on the individual's claim to have managed their own success. The result was that many women who felt content in their work as homemakers and mothers now feel as oppressed by the feminist movement as they had been by the patriarchal structures of the past.

CONCLUSIONS

In spite of troubling problems within the movement, the religious feminist social vision is the most promising vision for the third millennium. The breach in relationships between the sexes, between male and female, and between the human creature and God is clear to any thoughtful person. Whether this situation arose because of an original male and female couple eating a forbidden fruit is beside the point. The point is that this alienation/sin exists. Any social vision that does not deal with these broken relationships is only partial. The problem with the fundamentalism vision is that while it may seek to resolve the tension between the human creature and God it ignores the tension between the human creature and nature, as well as the broken relationship between male and female. Liberationism, on the other hand, is beginning to deal with broken relationships between the sexes, as well as with the broken relationship with the divine. It fails, however, to touch on the breech in relationship between the human being and nature. In fact, in its call for economic success above all else, it sometimes increases the break and treats nature in an exploitative fashion. Only feminism, at its best, deals with all three problems.

I realize at the end of this chapter that I have been dancing around the feminist social vision, never fully explaining it. Let me close this chapter with a comment by Elizabeth Moltmann-Wendel who, along with her husband, the theologian Jürgen Moltmann, participated in a World Council of Churches consultation on the Community of Men and Women in the Church, in Sheffield, United Kingdom, in June 1981. In the exchange between husband and wife, Elizabeth sketched out the feminist social vision with an aching beauty.

What women want is a new community in which those with power begin to listen to those without power. A community where there are opportunities for the powerless to express themselves and get organized. A community in which power is redistributed and those in power learn to give up their power – for the sake of justice. Women want a community which is not obsessed with profit and economic growth but concerned with the basic needs of all human beings...

What women want is a whole life, one which embraces body, soul, and spirit, no longer compartmentalized into private and public spheres; a life, moreover, which fills them with a trust and hope transcending biological death.[23]

To this vision I can only reply, "amen and amen."

CONCLUSION

THE LANDSCAPE OF THE FUTURE

It "is impossible to describe visions that will reflect conditions that have not yet come to pass," warns the philosopher Robert Heilbroner in his book *Visions of the Future*.[1] Hielbroner's caution is helpful as we try to gather together the threads of the previous chapters in order to sketch the contours of religious faith in the third millennium. I remember reading a book on the future written by Nicky Cruz, a former gang leader who became an evangelical Christian. Since I had enjoyed a previous book *The Cross and the Switchblade*, I anticipated great things from this sequel. Nicky Cruz did not disappoint; he listed event after event which he asserted the Bible predicted, often with quite specific details. Moreover, rather than leaving the dating open-ended, he claimed that all the events he wrote about would happen in the decade of the 1970s. The only trouble was that I was reading the book in 1980 when most of the events had failed to take place, which explained why it was on the bargain table selling at one-tenth of its original price![2]

With this caution in mind, then, as well as the admission that while an active Christian minister I am "neither a prophet, nor the son of a prophet"[3] let me sketch out some ideas of what the future of religion might be in North America and Europe.

AN EXPLOSION OF BELIEF

In their popular bestseller *Megatrends 2000*, authors John Naisbitt and Patricia Aburdene note that "at the dawn of the third millennium there are unmistakable signs of a worldwide multidenominational revival." As evidence they point to the emergence of the New Age movement, the fact that in 1987 the Mormons gained 274,000 new converts, the revival of Shinto neighborhood festivals in Japan, the explosive growth of the charismatic movement, and the interest which Chinese and Soviet young people are showing in religion, much to the dismay of their parents.[4]

The engine behind this growth in religious sensibility and commitment, according to Naisbitt and Aburdene, is the approach of the year 2000, with all its millennial symbolism. This is partially but not completely true. Although the turn of the millennium will have a profound effect on the religious psyche of many Westerners, nonetheless, it is not the main reason why there is such a widespread renewal of interest in religion.

The reason why this interest has surfaced has much more to do with the fact that, as St. Augustine put it, we are religious beings and have a spiritual hunger which cannot be satisfied by material things. In Augustine's words, "You [God] have made us for yourself, and our heart is restless until it rests in you."[5] Or, as Mother Henry says to Richard in James Baldwin's book *The Blues for Mister Charlie*, when he claims that he doesn't believe in God, "You don't know what you are talking about. Ain't no way possible for you not to believe in God. It ain't up to you." "Who is it up to then?" Richard asks. "It's up to the life in you – the life in you," Mother Henry responds. "That knows where it comes from, that believes in God. You doubt me, you just try holding your breath long enough to die."[6]

We are coming to the end of a period of history which has attempted to portray the human creature as a mere bundle of rags and bones and has confined the meaning of life to the accumulation of material goods. This period of history was really an anomaly all along, a rejection of much of what the ancients knew about life and its meaning. To be sure there were some benefits. The level of technology which we enjoy, the standard of living which we possess in the West are not to be scorned. However, it was not enough. It was never enough. As the singer Bruce Cockburn puts it, "there must be more." This search for *more* is what ignites the spiritual explosion which we see happening all around us.

At a wedding I conducted, my wife Cathy and I ended up at the reception that followed seated next to one of North America's bestselling novelists. Writing under the pen name Joy Fielding, the woman, who I later found out came from a Jewish background, turned to me in the middle of the dinner conversation and said wistfully, "You really do believe in God, don't you." When I replied that I did she said with pathos, "Oh, how I wish that I could." I wish I had had Mother Henry's wisdom at that moment and replied to her gently but firmly that believing was not up to her but rather up to the life spirit in her, which believed even in the midst of her unbelief.

Not, I hasten to add, that this revival of religious belief and spiritual awareness will be a tranquil affair. I can still recall an executive from a large multinational corporation making a special appointment to meet with me concerning problems he was having at work. He was in charge of personnel and was trying to figure out a way to keep religious convictions out of the workplace. He wondered if there was any advice I could give which would allow him, during the hiring process, to quietly filter out people with religious beliefs who might cause problems later. When I questioned him more, I found out that the particular problem he was dealing with concerned several Jehovah's Witnesses whose beliefs were conflicting with some of the company's activities. I replied that, while I sympathized with his concern, I saw things getting worse from his perspective rather than better.

While in some ways societies which separate the institutions of church and state are better able to accommodate the range of religious beliefs which will come bubbling to the surface, in other ways they are not. Religious beliefs are not a topic of study in our public schools. With attendance at religious institutions hovering around 30–40 percent, and with the vast majority of attendees over 40 years of age, many young people are unable to differentiate between the healthy expressions of spirituality which will arise and the unhealthy ones. The danger in this is that the spiritual hunger which is part of what it means to be human puts people at risk to the machinations of religious and cult leaders who will use that spiritual hunger for selfish gains rather than for the good of the individual involved or for the wider society.

However, because I trust that the Spirit of God is active and powerful, I suspect that most of the expressions of spirituality which will surface in the future will be healthy ones. At heart, I am optimistic that given enough time and enough information people will choose life-giving rather than life-denying options. The forms will be different but the common yearning for a relationship with the divine – the ground of our being, as the theologian Paul Tillich liked to phrase it – will propel the Western world in a much needed positive direction.

RAPPROCHEMENT BETWEEN SCIENCE AND RELIGION

Much of the impetus for this renewal of religion will result from the new world views being offered to us, due to the discoveries associated with Einstein's general and special theory of relativity, and with quantum mechanics. The computer chip, which is based on quantum theory, has already had a tremendous impact on our lives. For one thing, it is displacing traditional sources of wealth. The richest individual in the world is Bill Gates, the founder and majority shareholder of the Microsoft Corporation.[7] Already this fact is causing headaches for economists. How does one measure the value of ideas, which is basically what the Microsoft Corporation is? The IBM Company made this mistake when they allowed Bill Gates to retain ownership of the operating language he devised for their early personal computers. They reasoned that the most important item was the physical computer while Bill Gates reasoned that a box without software is just a box. In the end, he was right.

Moreover, with the advent of the Internet, time and space are changed in a way that, in the end, will have far greater impact than did the computer by itself. Alvin Toffler's "third wave," in which urbanization is reversed, finally becomes a possibility. According to Toffler, western civilization can be divided into three stages – " a First Wave agricultural phase, a Second

Wave industrial phase, and a Third Wave phase now beginning." [8] One of the most significant aspects of this third wave is the decentralization and deurbanization of society due to the computer revolution. It is, though, a reversal of urbanization which at the same time results in both a return to rural living and a new reality – cyberspace – which in effect transforms the entire world into one large urban center. I am an example of this, living in rural Nova Scotia but being part of the new cyberworld as owner of a company called *Church Online!* By means of the Internet, I am in touch with churches and church leaders throughout the world whom I have never met personally and likely never will meet, but with whom I communicate directly and instantly. Moreover, I buy books online, order software online, research information online, and my children chat online to strangers they have never met before but who, for a few days at least, become fast friends, sharing common interests. Even while living in urban centers I never felt as urbanized as I do now that I am residing in the country!

Recently, a reporter for a Christian *e-zine* (electronic magazine) telephoned to interview me about how I thought the Internet could be used by churches. She was particularly interested in the cyber-churches which I mentioned were active on the Internet, even though I myself am opposed to them. It is entirely possible (and in fact already is a present reality for some) that many people will find their religious affiliation not with a local church or synagogue but with a cyber one. This will add to the religious pluralism which surrounds us. For example, anyone who wants to be a part of a Wiccan community in rural United States and Canada might have a difficult time. However, with the Internet, all one need do is turn on the computer and log on in order to find a smorgasbord of sites which explain and support Wiccan beliefs.

The computer, along with its more important extension, the Internet, has had an enormous impact upon our life. But I predict that it is the impact which quantum mechanics and relativity theory have had and will have, particularly on our view of the cosmos, that will have the greatest influence upon our religious beliefs. An example of this is the attempt of scientists to map the topography of the universe. The ancients thought the world was flat and was the center of all that was. This belief persisted until Copernicus convinced us that the Earth revolved around the sun, rather than the sun rotating across the vault of heaven that covered the Earth. Suddenly, not only the Earth but also the human creature was displaced. This devolution of human importance became even more pronounced when scientists discovered that our sun was simply one of many and our galaxy one of a seemingly infinite number of galaxies. Now, not only was the Earth displaced but so was the sun. Every point in space was at the same time equally important and equally meaningless.

With Einstein's discovery that space is curved, as well as with the technology of microwaves which allows scientists to picture the growth of the universe from the Big Bang to the present, we are beginning to think that the universe may have a center once again. What is at this center and what this will mean philosophically and theologically is yet unknown. But that this discovery will have an impact on our thinking about life and the meaning of life is beyond question.

It is clear, then, that the religious groups who are best able to integrate the new perspectives arising out of quantum physics and relativity theory will be the groups that will lead the way into the future. Specifically, the new perspectives I am referring to are the importance of a continuing dialogue and partnership between science and religion; the impact of Heisenberg's Uncertainty Principle, one of the mainstays of quantum theory; and the implications of "spooky action at a distance" whereby two electrons which were once in union continue to interact with each other even though they may be sent in completely opposite directions. Added to this is the influence which quantum physics has upon cosmology through the Big Bang theory of the origin of the universe,[9] as well as the insights of the two theories of relativity and the continued attempt to reconcile these with quantum physics. These discoveries, and others, push us in the direction of holism, mystery, humility, and a growing recognition that we human creatures participate in the unfolding of the cosmos and are not just passive spectators.

What this means for the Christian community in the West is that the mainline church's compromise wherein faith is divorced from science, as well as the fundamentalist option which ignores the challenges of science are both inadequate options. To be precise, fundamentalism does not really ignore science, it simply freezes science at a certain point and takes selectively from it. In this partnership with science, fundamentalists are one step closer than many mainline Christians who divorce faith from science so completely that not only is confrontation not a possibility but neither is dialogue. At the same time, however, they are further away. Fundamentalists, at least of the North American variety, believe that science and religion belong together and that science supports religious beliefs. However, the scientific view they adopt is a Cartesian one interpreted through the grid of Thomas Reid's philosophy who, in opposition to the skepticism of Hume, insisted on the viability of common sense. Thus, fundamentalists are one step closer than mainline Christians in that they support the reunion of science and religion – one of the most promising and distinctive features of the religious landscape of the future. On the other hand, they are much further away in that the new world that quantum physics reveals to us is one full of surprises. As the physicist and theologian, John Polkinghorne puts it, the new scientific insights teach us that, "common sense is not the measure of everything."[10]

The impact that the new science will have on the conservative wing of Christianity will be the passing of leadership and influence from the fundamentalists and their kissing cousins, the evangelicals, to the neo-Pentecostals who continue to give lip service to fundamentalist doctrines but are already putting into practice some of the viewpoints which arise from new scientific views.

A prime example of how insights connected with the new views of physics are being incorporated in a practical manner into the religious beliefs and practices of the neo-Pentecostal churches is found in the arena of alternative medicine. While mainline Christians often scoff at the false hope engendered by these churches and their healing services, and fundamentalists complain that miracles ceased after the time of the apostles, people who suffer from illnesses, being less critical, flock to Pentecostal-type healing services. Moreover, the New Age movement, which also stresses alternative medical practices, continues to grow in popularity and public acceptance.

Another example concerns the fascination with life after death which has always been a part of human hopes but has surfaced in a dramatic way through the increased reports of near-death experiences. Recently, the moderator of the United Church of Canada, the largest mainline church body in that country, announced that he did not know what to make of the Christian doctrine of the resurrection of the dead. His comments, while causing a furor, including an attempt by a lawyer in Halifax to remove him from his post, are not very different from those made by some Anglican bishops in Great Britain or those found in the writings of various theologians.

Such comments, however, do not speak to the human heart nor, more to the point, do they reflect some of the insights arising out of the new physics. To be sure, the data is ambiguous and it may well prove that science will finally, to use Stephen Hawking's phrase, "know the mind of God" and thus dispose of religious faith entirely. However, it is just as likely that the reverse will happen. As we shift from viewing the human body as simply a biochemical machine to a vibrating energy field then, suddenly, alternative medical practices make not only religious but also scientific sense. Moreover, although NDE researchers are often treated as the lunatic fringe of the medical community, the evidence they are starting to accumulate is impressive, to say the least.

Personally, I have often speculated that since nothing we can see travels faster than light (according to Einstein who used light as his constant in both the theories of general and special relativity), it may well be possible that alternative healing practices, specifically prayer, manipulate an energy field which does move faster than light. If true, this means that some alternative medical practices at least are not anti-scientific but merely supra-scientific. Moreover, the same could be said of the resurrected body referred to in the apostle Paul's epistles. Is it simply a case that the dead have become supercharged, their molecules moving so quickly that they surpass the speed of light and are therefore invisible to us?

Such questions cannot yet be answered. The important point is they are no longer nonsensical questions but questions which scientists are beginning to study and to reflect on. Indeed, as the mathematician Frank Tipler notes after attending a theological conference in New Orleans, there exists the ironic situation today where mainline Christian scholars poke fun at talk of an afterlife at the very same time as scientists such as Tipler seek to prove its mathematical probability![11]

"SPIRITUALITY, YES. ORGANIZED RELIGION, NO"

One of the features of the religious landscape of the future, as has been abundantly evident in almost all the movements which have been examined, is a continuing rejection of institutionalized religion. Addressing the pluralizing and secularizing influences of the modern age, Richard Tarnas notes that, "while in most respects the influence of institutionalized religion has continued to decline, the religious sensibility itself seems to have been revitalized by the newly ambiguous circumstances of the postmodern era."[12]

To be sure, there will be a reaction and some people will be attracted to the most traditional forms of institutionalized faith. A few monasteries which were once on the decline have found new life. Ancient Gregorian chants have become popular once again. There is a continued exodus of evangelical ministers in the United States and Canada to more liturgical churches, initially the Episcopalian/Anglican Church and, more recently, various branches of the Orthodox Church. The vast majority of people in North America and Europe, however, will walk away from institutionalized faith or supplement their involvement in their chosen institution with non-institutionalized practices. As Naisbitt and Aburdene put it, "Spirituality, yes. Organized religion, no."[13]

This anti-institutionalism will be a doubled-edged sword. At some point in time, for example, the New Age groups who have attracted members from the mainline churches or the revivalist groups who have attracted members from conservative Christian churches will face the pressure to institutionalize themselves as well. What will happen to such groups? Will they be able to maintain their flexibility and spontaneity? This question is not easily answered. It may well be that the rejection of institutionalized faith will pose as serious a challenge to the new religious movements in the years ahead as it already does to established religious groups.

One thing is certain, religious institutions of the future will have to recognize and affirm pluralistic authorities rather than simply hierarchical or past authorities. Little wonder that the incisive and articulate theologian, Douglas Hall concludes his book *Thinking the Faith* by highlighting the impact that shifting views on religious authority will have on the Christian community. He notes, "there is thus a suggestion in our present situation that the most visible changes in the diaspora church that is coming to be will be those relating to the nature of ecclesial authority."[14] When he made this statement, Doug Hall must have had in mind the confrontations between Latin American liberationists and the Pope as well as the controversy between some North American bishops and the Vatican concerning the ordination of women to the priesthood.

The rejection of religious authority, however, will be much greater than simply concerns over the nature of ecclesial authority – over who runs the religious institutions. It will touch all levels of authority both inside and outside religious institutions. Religious groups that understand and cater to this anti-authoritarianism will grow while those who do not will either diminish or be forced into cult-like practices and behavior in order to retain support.

Much of this rejection of authority arises from the growth of religious pluralism which puts forward various, competing authorities in place of a common, unified authority. I grew up with the admonition, "God said it, the Bible records it, that settles it." In an age of religious pluralism, however, the lines of authority are not nearly as clear-cut nor as compelling. Part of this is due to the shrinking world in which we live. In the United States, adherents of the Muslim faith now outnumber Episcopalians; in Canada they outnumber Presbyterians. Moreover, the influence on Western society of Eastern religions, particularly Hinduism and Buddhism, has been immense, not only in regard to religious beliefs and practices but also in connection with popular culture as evidenced by the recent movie *Seven Years in Tibet*.

Granted, this influence has not always been a one-way street. I remember being accosted by a Hare Krishna devotee while standing in line waiting to get into a movie theatre. We started talking and he told me that there was much that was wrong with the world and with people. I agreed with him and then asked him if he knew one human being who had done no wrong; to use traditional Christian language, one who was without sin. Quickly, and to my surprise, he replied "Jesus." I agreed with him and the conversation ended there. Since in all likelihood the Hare Krishna devotee who accosted me was himself a North American by birth, a more compelling example of the influence of West on East, is the establishment and phenomenal growth of Christian churches in the country of South Korea. At one time the largest of these was a Presbyterian church which claimed 25,000 members. This has since been eclipsed by a Pentecostal church which claims 800,000 members – more than many denominations in both North America and Europe.

However, much of the influence has flowed East to West rather than the reverse. Indeed, the well-known sociologist Peter Berger claims that there have been three "contestations" in Christianity which "involved intensive and sustained intellectual effort" – the interaction of Christianity with the Greco-Roman world during the New Testament period, with Islam, and with modernity. He continues, "I believe we are now on the edge of a fourth weighty contestation, this time with the religious traditions of southern and eastern Asia."[15]

SEGMENTED AS AN ORANGE
– THE YEARNING FOR WHOLENESS

The president of our local university supposedly has threatened the teaching staff that he will not listen to any more submissions concerning academic matters containing the word "holistic." I sympathize with his position but can find no other word to describe the sense of fragmentation which many people experience, and the consequent thirst for wholeness which they express. For many people, as mentioned in the introduction, life feels as segmented as an orange, held together by the mere rind of one's will.

Consequently, many people today long for a sense of wholeness. In part, this is what is fueling the revival in spirituality. The philosopher Hegel once noted: "Religion is not consciousness of this or that truth in individual objects but of the absolute truth, of truth as the Universal, the all-comprehending, outside of which there lies nothing at all."[16] Sadly, in North America at least, even religion has contributed to this segmenting of truth since religious truth is often depicted as a specialized type of truth different from any other. Thus, while I personally welcome a growing interest in spirituality, I worry that too often the spiritual is being defined as simply one more component among many rather than as the living center to which all else coheres.

This is why I feel that the religious feminist vision, or a least certain expressions of it, is so important. At its best, religious feminism pushes toward wholeness and against the dualisms which have been culpable for much of the modern sense of fragmentation. While religious feminism keeps flirting with forming new dualisms every bit as distressing as those of the past, most religious feminists reject the body-spirit and the subject-object dualisms which have been important factors in contributing to the modern malaise. Likewise, Sam Keen in his book *Hymns to an Unknown God* notes, "the separation of spirit and flesh that pervades our culture confuses us all the more because deep down we know it is wrong. Each of us has a drive toward integration and wholeness,..."[17]

The problem, of course, is that we do not know how to combine unity with diversity. Feminists have rightly pointed out that much of what passed for unity in prior eras was really the domination of white males and the subjection of everyone else. In reaction, we have experienced the reign of relativism where "my truth is every bit as good as yours." This may well be a necessary development in the understanding of truth but it cannot be the final stage. The fundamentalist stance – that truth is obvious and clear and that only the pigheaded would disagree – is no longer sustainable. But, the opposite reaction that truth is infinitely malleable also has problems. For myself, I yearn for religious communities where truth is acknowledged but where a pluralism of viewpoint is celebrated in the search for and acceptance of that truth. I think such communities will fare well in the next century.

Typical of the type of communities which I am referring to are First Baptist, Kingston, Ontario, where I once had the privilege of working, and my present church, the Pereaux United Baptist Church in Nova Scotia. First Baptist, Kingston, is located in a city which is home to many of Canada's federal prisons, as well as one of Canada's most prestigious universities. One baptismal service in particular stands out. The two candidates for baptism were a son of the vice-president of Queens University and a former convict named Charlie, who later became a good friend. What made the service so special was that, apart from the minister, no one really thought it was all that special. This was what the church community was supposed to be – a community of faith in which distinctions of class, wealth, and education did not divide. In my present church, this same sense of diversity in unity is evident in the adult Sunday school class where conservatives and liberals challenge each other and disagree with each other, all in the context of a community of love. In part, this is due to the fact that my present church is located in a rural context where we all have to live together but I think that it is also more than that – it is the recognition that the divine binds us together *because* of our differences and not in spite of them.

THE CHRISTIAN OF THE
FUTURE MUST BE A MYSTIC

In his delightful book *God's Dominion: a Skeptic's Quest*, Ron Graham cites one of the most important theologians within the Roman Catholic church, the late Karl Rahner, as claiming that "the Christian of the future will be a mystic, or he or she will not exist at all."[18] I agree wholeheartedly with this prediction, which I also view as a prescription. The emphasis on correct doctrine so evident among conservative Christians, along with the emphasis on correct action which is the hallmark of liberal Christians, must play second fiddle to the emphasis upon a mystical oneness with the divine. It is only from this sense of oneness that doctrine and action can become alive. Indeed, without mysticism both doctrine and action lapse into prisons of the spirit. As the apostle James put it in his celebrated comment, "even the demons believe."[19] The missionary doctor and theologian, Albert Schweitzer, writing in the first half of this century, recognized the problem of an activism divorced from mysticism when he noted:

The spirit of the age drives us into action without allowing us to attain any clear view of the objective world and of life. It claims our toil inexorably in the service of this or that end, this or that achievement. It keeps us in a sort of intoxication of activity so that we may never have time to reflect and to ask ourselves what this restless sacrifice of ourselves to ends and achievements really has to do with the meaning of the world and of our lives. And so we wander hither and thither in the gathering dusk formed by lack of any definite theory of the universe like homeless, drunken mercenaries, and enlist indifferently in the service of the common and the great without distinguishing between them.[20]

The key role that mysticism will play, as already alluded to in the chapter on the New Age movement, is the answer to the emphasis on individualism which has been characteristic of the modern era. There is a growing appreciation that individualism is not the answer, that it must be balanced with a stress on responsibility and community. Some members of various cult groups or fundamentalist churches which cluster around charismatic, authoritarian leaders, have given up their individual freedom because it has proved too confusing and too painful. The vast majority of people, however, jealously guard their freedom of belief and practice but do so with a growing sense of unease.

We in the West are finally beginning to outgrow the adolescent rebellion of the Enlightenment, which caused us to focus on the authority of the "self." This presents both a danger and an opportunity. My father, a retired minister, feels that it presents a danger as people are increasingly drawn toward fundamentalist options, particularly Muslim fundamentalism. He is worried that the future will lie with militant groups who exchange individual freedom for group certainty and control. I think that he is misreading what is going on, even though I grudgingly admit that his vision is possible.

As I see it, the "foreign" authority of the church in the West was displaced, following the Reformation, with the authority of the self. This autonomy, where one is a law unto oneself, reached a high-water mark in our century but it is now beginning to show its age. As it does so, the opportunity opens to a new authority, not the adolescent tyranny of the self, but rather the authority of the Spirit, deep within.

In this, I am simply parroting the theologian Paul Tillich who saw us moving through *heteronomy*, to *autonomy*, and finally to *theonomy*. Tillich applied this concept not only to individuals but to cultures as well, noting that a "theonomous culture is spirit-determined, and Spirit-directed." It is a culture where "Spirit fulfils spirit instead of breaking it."[21]

Tillich's observations, although couched in difficult words, are obvious to any mother or father. The child, under parental authority and direction for the first years of his or her life, believes what the parent believes and sees the world through the parent's spectacles. In time, however, the child grows to be an adolescent and rebels. Parents will no longer be the foreign powers which dominate life and thoughts; instead the teen will be his or her own authority. This autonomous stance can last until death with some individuals, but for others who realize that the self can be every bit as foreign a master as one's parents, a final step needs to be taken. This is the mystical step where the authority is not self, but Spirit speaking to spirit.

This third stage brings to mind the prophet Jeremiah's prediction of a new covenant which would be written not on tablets of stone, extraneous to the individual, but rather on the tissue of the human heart. As Jeremiah saw it, this new covenant would not be mediated through religious authorities but through the Spirit of God testifying to the spirit of the individual. In other words, not heteronomy, nor even autonomy, but rather theonomy.

Where does this stress on theonomy, which can only be mediated through a mystical understanding of faith, leave Christianity, still the dominant faith of Western culture? The answer is that it leaves Christianity with a challenge. Mainstream interpretations of Christianity have focused on either correct doctrine or correct action as encompassing the heart of the Christian faith. Little wonder that many, myself included have despaired of the institutional Christian church and the faith it presents. However, driven in part by Eastern religious insights as well as by a rediscovery of the rich mystical traditions of Christianity, the divine Spirit is pushing our society toward spiritual maturity. Many in the Christian religion and beyond are beginning to discover a third option to activism or doctrinalism/rationalism – the option of mysticism.

When I reflect on my own experience, I think that I have always known at heart that Christian faith is mystical at its inner core, although I have often forgotten to live and worship by this insight. I remember one incident which happened to me while I was around twelve years old. I was sitting on a long wooden bench in a mud hut which served as a church in the poorer sections of the city of La Paz, Bolivia. The room was small and my back ached since there was no back to the bench and Bolivian Baptist church services often went on for hours at a time. Moreover, it stank of sweat and other odors as the Bolivians in that community had no access to showers or baths. I was singing as best I could in the Aymara language about the love of God for us as revealed in the person of Jesus, when I felt a sensation like an electric current shoot through my body. At first I dismissed this "feeling" until it happened another Sunday and then one more time.

I never shared this with anyone because I myself could not make sense of it at the time. Now I think I know what it was. It was the mystical sense, for only a fleeting moment, of oneness with God, a sense deeper than mere emotionalism and far richer than any theological doctrine. Perhaps the poet T. S. Eliot was right all along when he noted that "the end of all our exploring will be to arrive where we started and know the place for the first time." [22]

ENDNOTES

INTRODUCTION

[1] Robert Lewis, "From the Editor," *Macleans,* July 7, 1996, p. 2.

[2] The term "third-wave" has been used by writers such as Peter Wagner. According to this topology, the first wave of Pentecostalism began on New Year's Day, 1901 at a small Bible School in Topeka, Kansas, when a student named Agnes Ozman began to speak in strange sounds commonly known as "tongues" or more technically as "glossolalia." The second wave corresponds to the rise of the charismatic movement in the 1960s within the Roman Catholic, Anglican, and Lutheran traditions. The third wave of Pentecostalism is associated with such groups as the Vineyard Church and more recently, the Toronto Airport Christian Fellowship. This third wave has penetrated denominational groups that were doctrinally resistant to Pentecostalism, such as the Baptists, Presbyterians, and Christian Reformed.

[3] See Douglas John Hall, *Thinking the Faith: Christian Theology in a North American Context* (Minneapolis: Fortress Press, 1989).

[4] See, Mary Daly, *Pure Lust: Elemental Feminist Philosophy* (Boston: Beacon Press, 1984).

[5] Tillich argues that what concerns men and women ultimately is God for them and thus everyone is religious since everyone has ultimate concerns. "*God* is the answer to the question implied in man's finitude; he is the name for that which concerns man ultimately. This does not mean that first there is a being called God and then the demand that man should be ultimately concerned about him. It means that whatever concerns a man ultimately becomes god for him, and conversely, it means that a man can be concerned ultimately only about that which is god for him." Paul Tillich, *Systematic Theology,* Vol. 1 (Chicago: University of Chicago Press, 1951), p. 211.

[6] Paul Davies, *God and the New Physics* (New York: Simon and Schuster, 1983).

[7] Malcolm W. Browne, "Scientists Deplore Flight of Reason," *New York Times,* June 6, 1995.

[8] Frank J. Tipler, *The Physics of Immortality: Modern Cosmology, God and the Resurrection of the Dead* (New York: Anchor Books, 1994), p. 339.

[9] See Fritjof Capra, *The Tao of Physics* (London: Fontana, 1975).

CHAPTER ONE

[1] See Mari Naumvoski, "Strengthening Vision: Do Glasses Protect Us From the World?" *Healing Journeys: (Taking Control)*. Susan Drake and Carolyn Dailey Klopstock eds. (South Surrey, British Columbia: Pinecrest Press, 1992).

[2] Tom Harpur, *The Uncommon Touch: an Investigation of Spiritual Healing* (Toronto: McClelland and Stewart, 1995), p. 214.

[3] David Spiegel, *Living Beyond Limits* (New York: Ballantine Books, 1993), p. 227.

[4] Chris Haftner-Eaton and Laurie Pearce, "Birth Choices, the Law, and Medicine: Balancing Individual Freedoms and Protection of the Public's Health," *Journal of Health Politics, Policy and Law* 19 (Winter 1994), p. 831.

[5] Richard A. Knox, "Agency on Medical Cost-Effectiveness Fighting for Life," *Boston Globe* (23 July 1995), p. 16.

[6] Larry Dossey, *Healing Words: the Power of Prayer and the Practice of Medicine* (San Francisco: Harper San Francisco, 1993), p. xix.

[7] Fred Frohock claims that four features are common to all forms of Pentecostalism: "salvation through conversion, baptism in the Holy Spirit, an expectation that Christ will come to earth a second time, *and a belief in divine healing through faith.*" See Fred M. Frohock, *Healing Powers: Alternative Medicine, Spiritual Communities, and the State* (Chicago: University of Chicago Press, 1992), p. 113.

[8] Mary Farrell Bednarowski, "Holistic Healing in the New Age," *Second Opinion* 19 (January 1994), p. 64.

[9] "Report on Needs Assessment for Primary Health Care in Eastern Kings," submitted by Verna Powell and Joanna M. DeLong (October, 1994).

[10] Advocates of alternative medicine prefer the term "allopathic" rather than mainstream or traditional medicine in order to avoid the pejorative connotation created by the contrast between the terms mainstream and alternative. As opposed to homeopathic treatment, allopathic treatments treat conditions/diseases with their opposites: i.e. antispasmodics, antidepressants etc.

[11] Phillip R. Alper, "Avoiding the *Marginal* Magic," *The Health Robbers:a Close Look at Quackery in America*, edited by Stephen Barrett and William T. Jarvis (Buffalo: Prometheus Books, 1993), p. 125.

[12] Harpur, *The Uncommon Touch*, p. 105.

[13] Larry Dossey, *Healing Words*, p. 6.

[14] Peter Duffy, "Faith Healers: Lightning Rods to our Own Inner Power?" *The Chronicle-Herald* (August 27, 1996), p. A7.

[15] See David Speigel, *Living Beyond Limits*, p. 71.

[16] John Langone, "Challenging the Main Stream," *Time Special Issue: the Frontiers of Medicine* (Fall, 1996), p. 32.

[17] See *New England Journal of Medicine* (January 28, 1993); and Patricia Chisholm, "Healers or Quacks," *Maclean's* (25 September 1995), p. 34.

[18] Philip Clark and Mary Jo Clark, "Therapeutic Touch: Is there a Scientific Basis for the Practice?" *Examining Holistic Medicine*, edited by Douglas Stalker and Clark Clymour (Buffalo, New York: Prometheus Books, 1985), p. 294.

[19] The journalist-priest Tom Harpur gives a very readable account of Grad's work with Estebany in his book *The Uncommon Touch*.

[20] Robert C. Fuller, "The Turn to Alternative Medicine," *Second Opinion*, 18 (July 1992), p. 11.

[21] Dossey, *Healing Words*, p. 205.

[22] See my discussion on the millenarianistic background to the New Age movement in chapter four.

CHAPTER TWO

[1] This was one of the terms originally chosen by Lovelock for his new theory. See Theodore Roszak, *The Voice of the Earth* (New York: Simon and Schuster, 1992), p. 155.

[2] James Lovelock, *Gaia: a New Look at Life on Earth* (Oxford: Oxford University Press, 1979), p. 10.

[3] Lovelock, *Gaia*, p. viii.

[4] James Lovelock, *"Gaia: A Model,"* in *Gaia a Way of Knowing: Political Implications of the New Biology* edited by William Irwin Thompson (Hudson, New York: Lindisfarne Press, 1987), p. 85.

[5] Lovelock, *Gaia*, p. 6.

[6] Elisabet Sahtouris, *Gaia: the Human Journey from Chaos to Cosmos* (New York: Pocket Books, 1989), p. 21.

[7] Lovelock, *Gaia*, p. 8.

[8] Lovelock, *Gaia*, p. 9.

[9] James Lovelock, "Gaia," in *Gaia 2 Emergence: the New Science of Becoming*, edited by William Irwin Thompson (Hudson, New York: Lindisfarne Press, 1991), p. 30.

[10] Lovelock, "Gaia," *Gaia 2 Emergence: the New Science of Becoming*, p. 31.

[11] Sahtouris, *Gaia: the Human Journey from Chaos to Cosmos*, pp. 63-64.

[12] William Irwin Thompson, "Symposium: from Biology to Cognitive Science," in *Gaia 2 Emergence: the New Science of Becoming*, p. 224.

[13] Lovelock, *Gaia*, pp. ix-x.

[14] Loren Wilkinson, "Gaia Spirituality: a Christian Critique," *Themelios*, vol. 18, no. 3 (April 1993), p. 4.

[15] Lovelock, *Gaia*, p. 12.

[16] Lynn Margulis, "Early Life," *Gaia a Way of Knowing: Political Implications of the New Biology*, p. 109.

[17] T. S. Eliot, *Collected Poems 1909-1962* (London: Faber and Faber, 1963), p. 92.

[18] Roszak, *The Voice of the Earth*, p. 189.

[19] Alan S. Miller, *Gaia Connections* (Savage, Maryland: Rowman and Little, 1991), p. 18.

[20] Roszak, *The Voice of the Earth*, p. 127.

[21] Rosemary Radford Ruether, *Gaia and God: an Ecofeminist Theology of Earth Healing* (New York: HarperCollins, 1992), pp. 36-37.

[22] Miller, *Gaian Connections*, p. 119.

[23] Sahtouris, *Gaia: the Human Journey from Chaos to Cosmos*, p. 34.

[24] James Lovelock, "Foreword," in Sahtouris, *Gaia: the Human Journey from Chaos to Cosmos*, p. 14.

[25] Cited in Tod Connor, "Is the Earth Alive?" *Christianity Today* (January 11, 1993), p. 24.

[26] "Godliness and Greenness," *The Economist* (December 21, 1996), p. 108.

[27] Ian L. McHarg, *Design In Nature* (Garden City, New York: Natural History Press, 1969), p. 24.

[28] Thompson, "Introduction," *Gaia: the Science of Becoming*, p. 25.

[29] Sahtouris, *Gaia: the Human Journey from Chaos to Cosmos*, p. 28.

[30] Rosemary Radford Ruether, *Gaia and God: an Ecofeminist Theology of Earth Healing*, p. 152.

[31] Martin Palmer, "Dancing to Armageddon: Doomsday and Utopia in Contemporary Science and Religion," *CTNS Bulletin*, vol. 12, no. 2 (Winter, 1992), p. 6.

[32] Rosemary Radford Reuther, *Gaia and God: an Ecofeminist Theology of Earth Healing*, p. 252.

[33] Sam Keen, *Hymns to an Unknown God: Awakening the Spirit in Everyday Life* (New York: Bantam Books, 1994), p. 176.

[34] Roszak, *The Voice of the Earth*, p. 221.

CHAPTER THREE

[1] Stephen W. Hawking, *A Brief History of Time: from the Big Bang to Black Holes* (New York: Bantam Books, 1988), p. 175.

[2] "A Mystic Universe," *The New York Times*, January 28, 1928 cited in L. Pearce Williams, ed. *Relativity Theory: Its Origins and Impact on Modern Thought*, Major Issues in World History Series (New York: John Wiley and Sons, 1968), pp. 129-130.

[3] Roger Penrose, *The Emperor's New Mind: Concerning Computers, Minds, and the Laws of Physics* (Oxford: Oxford University Press, 1989), p. 228

[4] Hawking, *A Brief History of Time*, p. 9.

[5] David Lindley, *The End of Physics: the Myth of a Unified Theory* (New York: Basic Books, 1993), p. 255.

[6] Lindley, *The End of Physics*, p. 77.

[7] Gary Zukav, *The Dancing Wu Li Masters: an Overview of the New Physics* (Toronto: Bantam Books, 1979), p. xxvix.

[8] The details of Einstein's special theory of relativity are interesting but outside the scope of this chapter to explain. The best popularization, in my opinion, remains Bertrand Russell's little book entitled *ABC of Relativity*, now in a 4th edition and published by Unwin Paperbacks in London, England.

[9] Roger S. Jones, *Physics for the Rest of Us: Ten Basic Ideas of Twentieth-Century Physics That Everyone Should Know... and How They Have Shaped Our Culture and Consciousness* (Chicago: Contemporary Books, 1992), p. 159.

[10] Paul Davies, *God and the New Physics* (New York: Simon & Schuster, 1983), p. vii.

[11] Nancy Logan, "Godhead Revisited," *Toronto Life Fashion* (December/January 1993), p. 52.

[12] Robert Jastrow, *God and the Astronomers* (New York: W. W. Norton and Company, 1978), p. 116.

[13] Jones, *Physics for the Rest of Us*, p. 345.

[14] Francis Crick, *The Astonishing Hypothesis: the Scientific Search for the Soul* (London: Touchstone Books, 1994), p. 257.

[15] Frank Tipler, *The Physics of Immortality: Modern Cosmology, God and the Resurrection of the Dead* (New York: Anchor Books, 1994), p. 8.

[16] Werner Heisenberg, *Physics and Beyond: Encounters and Conversations*, translated by Arnold J. Pomerans (New York: Harper and Row, 1971), p. 11.

[17] Richard Tarnas, *The Passion of the Western Mind: Understanding the Ideas that Have Shaped our World View* (New York: Ballantine Books, 1991), p. 357.

[18] See Fritjof Capra and David Steindl-Rast, *Belonging to the Universe: Explorations on the Frontiers of Science and Spirituality* (New York: Harper Collins, 1991), pp. xi-xv. In this book, Fritjof Capra lists his "paradigm shift" in the field of science and beside it, David Steindl-Fast, a Benedictine monk, lists his corresponding shift in theology.

[19] See Marilyn Ferguson, *The Aquarian Conspiracy: Personal and Social Transformation in Our Times* (London: Paladin, 1980).

[20] Lindley, *The End of Physics*, p. 54.

[21] Ian Barbour, *Religion in an Age of Science*, Gifford Lectures, vol. 1 (San Francisco: Harper and Row, 1990), p. 122.

[22] Stanley Jaki, *The Relevance of Physics* (Chicago: University of Chicago Press, 1966), p. 436.

[23] Jaki, *The Relevance of Physics*, p. 451.

[24] Tipler, *The Physics of Immortality*, p. 5.

[25] Ralph Wendell Burhoe, "The Center for Advanced Study in Religion and Science, and Zygon: Journal of Religion and Science – a Twenty-Year Review," in *Zygon*, vol. 22, 1987, p. 4.

[26] Willem B. Drees, "Problems in Debates About Physics and Religion," in *Physics and Our View of the World*, edited by Jan Hilgevoord (Cambridge: Cambridge University Press, 1994), p. 208.

CHAPTER FOUR

[1] Cited in Henry Bettenson, ed., *The Early Christian Fathers: a Selection from the Writings of the Fathers from St. Clement to St. Athanasius* (London: Oxford University Press, 1976), p. 293. Of course, Athanasius who was elected bishop of Alexandria in 328, was immersed in a heated theological controversy over the humanity and divinity of Jesus and in order to preserve Jesus' divinity was compelled to make human beings divine as well. It would be unfair to take his comments out of their historical context and apply them to our own situation. Indeed, I suspect that since Athanasius was careful to avoid making God and Jesus identical, thus falling into the trap of Modalism whereby God the Father changes into God the Son, that he would also be careful to differentiate the deity of the human being from the primary deity of Jesus.

[2] Martin Palmer, *Coming of Age: an Exploration of Christianity and the New Age* (Hammersmith, England: Aquarian Press, 1993), p. 89.

[3] Palmer, *Coming of Age*, pp. 38-39.

[4] Wouter J. Hanengraaff, *New Age Religion and Western Culture: Esotericism in the Mirror of Secular Thought* (Leiden: E. J. Brill, 1996), p. 95.

[5] Cited in Michael York, The *Emerging Network: a Sociology of the New Age and Neo-Pagan Movements* (Lanham, Maryland: Rowman and Littlefield, 1995), p. 64.

[6] In this regard, it is interesting that one of the foremost historians of revival movements in the United States, William McLoughlin, has termed the counterculture revolution of the sixties as "the Fourth Great Awakening." According to McLoughlin, the counterculture movement was a reaction against the liberalism that was the primary feature of the Third Great Awakening, which he dates from the late 1800s to 1920. See William G. McLoughlin, *Revivals, Awakenings, and Reform: an Essay on Religion and Social Change in America, 1607-1977* (Chicago: University of Chicago Press, 1978), pp. 179-216.

[7] Cited in Marci MacDonald, "The New Spirituality," *Macleans* (October 19, 1994), p. 46.

[8] Ted Peters, *The Cosmic Self: a Penetrating Look at Today's New Age Movements* (San Francisco: Harper San Francisco, 1991), p.2.

9 Marilyn Ferguson, *The Aquarian Conspiracy: Personal and Social Transformation in Our Times* (London: Paladin, 1980).

10 See José Argüelles, *The Mayan Factor: Path beyond Technology* (Santa Fe, New Mexico: Bear and Company, 1987).

11 Riane Eisler, *The Chalice and the Blade* (London: Mandala, 1990), p. 203.

12 Stephen M. Gelber and Martin L. Cook, *Saving the Earth: the History of a Middle-Class Millenarian Movement* (Berkley: University of California Press, 1990), p. 98.

13 Elizabeth L. Hillstrom, *Testing the Spirits* (Downers Grove, Illinois: InterVarsity Press, 1995), p. 213.

14 Palmer, *Coming of Age*, p. 13.

15 Robert Lewis, "From the Editor," *MacLeans* (July 7, 1996). In a poll directed by my friend and former parishioner, the late George Rawlyk, and done in cooperation with the Angus Reid Group, *Macleans* was shocked to find that Canadians were almost as religious as their American counterparts. As the information blurb on the index page puts it, "a major new poll contradicts the conventional wisdom that Canadians have turned their backs on religion." See also "God Is Alive," *Macleans* (April 12, 1993), pp. 32-50.

16 Cited in Todd Deaton, "New Wineskins Needed for New Era," *Atlantic Baptist* (July-August, 1996).

17 Hillstrom, *Testing the Spirits*, p. 213.

18 Hanengraaff, *New Age Religion and Western Culture*, p. 524.

19 Henry Gordon, *Channeling in the New Age: the "Teachings" of Shirley MacLaine and Other Such Gurus* (Buffalo, New York: Prometheus Books, 1988), p. 67.

20 Michael Cole et al, *What is the New Age?* (London: Hodder and Stoughton, 1990), p. 95

21 Gerard Manley Hopkins, "God's Grandeur," *The Batsford Book of Religious Verse*, edited by Elizabeth Jennings (London: B.T. Batsford, 1981), p. 54.

22 Palmer, *Coming of Age*, p. 27.

23 Ian Harris, Stuart Mews, Paul Morris, John Shepherd, *Contemporary Religions: a World Guide* (Essex, England: Longman Current Affairs, 1992), p. 252.

24 Cole, *What is the New Age?*, p. 60.

25 M. D. Faber, *New Age Thinking: a Psychoanalytic Critique* (Ottawa: University of Ottawa Press, 1996), p. xiii.

26 Cited in Robert Basil, ed., "Introduction," *Not Necessarily the New Age* (Buffalo, New York: Prometheus Books, 1988), p. 20.

CHAPTER FIVE

1 Cottie A. Burland, *"Primitive Societies,"* in *Life after Death*, edited by Arthur Toynbee and Arthur Koestler (London: Weidenfeld & Nicolson, 1976), pp. 52-53.

[2] Cited in Karlis Osis and Erlendur Haraldsson, *At the Hour of Death* (New York: Avon Books, 1977), p. 15-16.

[3] W.F. Barrett, *Death-Bed Visions* (London: Methuen, 1926).

[4] Raymond Moody, *Life after Life* (New York: Mockingbird Books, 1975).

[5] Melvin Morse, *Transformed by the Light: the Powerful Effect of Near-Death Experiences on People's Lives* (New York: Ivy Books, 1992), pp. 22-23.

[6] Cherie Sutherland, *Within the Light* (New York: Bantam Books, 1995), p. 94.

[7] Margot Grey, *Return from Death: an Exploration of the Near-Death Experience* (London: Arkana, 1985), p. 72.

[8] Plato, *The Republic*, translated by Francis M. Cornford (London: Oxford University Press, 1945), p. 351.

[9] C. G. Jung, *Memories, Dreams, Reflections*, edited by Aniela Jaffé (New York: Vintage Books, 1965), p 289.

[10] Ibid, p. 292.

[11] See, for example, Jenny Randles and Peter Hough, *The Afterlife: an Investigation into the Mysteries of Life after Death* (New York: Berkley Books, 1993), pp. 218-220.

[12] Tom Harpur, *The God Question and Other Faith Issues* (Hantsport, Nova Scotia: Lancelot Press, 1993), p. 33.

[13] Malley Cox-Chapman, *The Case for Heaven: Near-Death Experiences as Evidence of the Afterlife* (New York: G. P. Putnam's Sons, 1995), p. 165.

[14] Angie Fenimore, *Beyond the Darkness: My Near-Death Journey to the Edge of Hell* (New York: Bantam Books, 1995), p. 96.

[15] Hans Küng, *Eternal Life? Life after Death as a Medical, Philosophical and Theological Problem*, translated by Edward Quinn (Garden City, New York: Doubleday and Company, 1984), p. 15.

[16] Susan Blackmore, *Dying to Live: Near-Death Experiences* (Buffalo, New York: Prometheus Books, 1993), p. 43.

[17] Carl Sagan, *Broca's Brain* (New York: Random House, 1979), p. 143.

[18] Rosalind Heywood, "Illusion – or What?" in *Life after Death*, edited by Arthur Toynbee and Arthur Koestler (London: Weidenfeld & Nicolson, 1976), p. 234.

[19] Morse, *Parting Visions: Uses and Meanings of Pre-Death, Psychic, and Spiritual Experiences*, introduction by Betty J. Eadie (New York HarperCollins, 1994), p. 64.

[20] Cited in Sutherland, *Within the Light*, p. 149.

[21] Ibid, pp. 49-50.

[22] Cited in Margot Grey, *Return from Death*, p. 108.

[23] Cited in Sutherland, *Within the Light*, p. 81.

[24] Betty Eadie, *Embraced by the Light* (New York: Bantam Books, 1992), p. 46.

[25] J. Bruce King, "Reincarnation," *Death, Afterlife, and the Soul*, edited by Lawrence E. Sullivan, Religion, History, and Culture Selections from the *Encyclopedia of Religion*, edited by Mircea Eliade (New York: Macmillan Publishing Company, 1989), p. 140.

[26] Cited in Sutherland, *Within the Light*, pp. 70-71.

[27] Ibid, p.26.

[28] See David Lorimer, *Whole in One: the Near-Death Experience and the Ethic of Interconnectedness* (London: Arkana, 1990).

[29] Harpur, *Life after Death* (Toronto: McClelland & Stewart, 1991), pp. 46-47.

[30] Cox-Chapman, *The Case for Heaven*, p. 192.

[31] Grey, *Return from Death*, p. 193.

[32] Cited in Harpur, *Life after Death*, p. 62

CHAPTER SIX

[1] Throughout this chapter, I will refer to the Toronto Airport Christian Fellowship as the Toronto Airport Church for the sake of simplicity.

[2] Harvey Cox, *Fire from Heaven: the Rise of Pentecostal Spirituality and the Reshaping of Religion in the Twenty-first Century* (Reading, Massachusetts: Addison-Wesley Publishing, 1995), pp. 64-65.

[3] Ibid, p. 139.

[4] John Wimber, "Spirit Song," in *Master Chorus Book: Contemporary, Traditional and New Choruses*, compiled by Ken Bible (Kansas City, Missouri: Lillenas Publishing, 1987).

[5] Wayne Grudem, *Power and Truth: a Response*, Vineyard Position Paper Number 4 (Anaheim, California: Association of Vineyard Churches, 1993), p. 65.

[6] James A. Beverley, *Holy Laughter and the Toronto Blessing: an Investigative Report* (Grand Rapids: Zondervan Publishing House, 1995), pp. 144-145.

[7] This juxtaposition of contrary images is a feature of these prophecies as water and fire are two images which are often used. The physical manifestations also seem to incorporate contrary elements. Arnott describes what happened to a man named Terry Virgo who experienced the descent of the Spirit. He writes: "Terry Virgo, the international overseer of New Frontiers Churches, wrote to me about his visit to our church. Earlier Carol and I were with him in Brighton, England, in October 1994. At one point he was lying down with one of us at his head and one at his feet, soaking him in the power of God. He was saying, 'My whole body is on fire. I have never been this high before. I don't know what to do.' When I asked him later, he said he was afraid, and 'My whole body was *on fire*. I felt there was a *driving hailstone* in my face.'" [emphasis mine]See John Arnott, *The Father's Blessing* (Orlando, Florida: Creation House, 1995), p. 80.

[8] Dave Roberts, *The 'Toronto' Blessing* (Eastbourne, England: Kingsway Publications, 1994), p. 16.

⁹ Arnott, *The Father's Blessing*, p. 58.

¹⁰ Robert Hough, "God is Alive and Well and Saving Souls on Dixie Road," in *Toronto Life* (February 1995), p. 31.

¹¹ Cited in Beverley, *Holy Laughter and the Toronto Blessing*, p. 54.

¹² Carol Arnott, *"Sword Prophecy,"* Awakenings List – awakening@listserver.com – moderated by Richard Riss (23 January 1997).

·¹³ Harvey Cox, *Fire from Heaven*, p. 300.

¹⁴ See Harvey Cox, *Religion in the Secular City: Toward a Postmodern Theology* (New York: Simon and Schuster, 1984).

¹⁵ William DeArteaga, *Quenching the Spirit: Discover the Real Spirit Behind the Charismatic Controversy* (Orlando, Florida: Creation House, 1996), p. 157.

¹⁶ Cited in Beverley, *Holy Laughter and the Toronto Blessing*, p. 89.

¹⁷ Cox, *Fire from Heaven*, p. 115.

¹⁸ *"Synopsis of the Week 20 May-1 June 1997,"* Internet Document (http://www.brownsville-revival.org/c_news.html)

¹⁹ Julie Suggs, "Revival in Valdosta!" Awakenings List (14 June 1997).

²⁰ Cited in Beverley, *Holy Laughter and the Toronto Blessing*, p. 61.

²¹ Cited in John Wimber and Kevin Springer, *Power Healing*, (San Francisco: HarperSanFrancisco, 1987), p. xx.

²² "Toronto Airport Church Fellowship Statement of Faith," Internet Document (http://www.tacf.org/revivalnews/statfaith.html).

²³ Cited in Guy Chevreau, *Catch the Fire* (London: HarperCollins, 1994), pp.. 146-149. For his investigation of alleged errors in this story see Beverley, *Holy Laughter and the Toronto Blessing*, pp. 103-120.

²⁴ Margaret Poloma, "By Their Fruits: a Sociological Analysis of the 'Toronto Blessing,'" (http://www.tacf.org/revivalnews/ mmpfruit.html), p. 7.

²⁵ "The Holy Spirit: God at Work," *Christianity Today* (March 19, 1990), pp. 27-35.

²⁶ Chuck Schmitt, *Awakening List* (June 21, 1997).

²⁷ Arnott, *The Father's Blessing*, p. 153.

²⁸ Laurie Barber, "Will We Miss the Blessing?" in *The Canadian Baptist* (March 1995), p. 12.

²⁹ See Arnott, *The Father's Blessing*, p. 181.

³⁰ Ibid, p. 87.

³¹ "He said, 'Go out and stand on the mountain before the Lord, for the Lord is about to pass by.' Now there was a great wind, so strong that it was splitting mountains and breaking rocks in pieces before the Lord, but the Lord was not in the wind; and after the wind an earthquake, but the Lord was not in the earthquake; and after the earthquake a fire, but the Lord was not in the fire; and after the fire a sound of sheer silence. When Elijah heard it, he wrapped his face in his mantle and went out and stood at the entrance of the cave. Then there came a voice to him that said, 'What are you doing here, Elijah?'" (1 Kings 19:11-13 NRSV)

CHAPTER SEVEN

[1] Stephen is not a real person but his religious conviction is real enough. In 1984, core members of the Gush Emunim joined a Jewish underground plot to blow up not only the Dome of the Rock but also the nearby al-Aska Mosque. If they had been successful one can only imagine the holy war which would have ensued. See Martin E. Marty and R. Scott Appleby, *The Glory and the Power: the Fundamentalist Challenge to the Modern World* (Boston: Beacon Press, 1992), p. 28.

[2] George A. Rawlyk, *Champions of the Truth: Fundamentalism, Modernism, and the Maritime Baptists* (Montreal and Kingston: McGill-Queen's University Press, 1990), p. 88.

[3] It was Riley who gave the evangelist Billy Graham his start by hiring him to superintend a group of independent schools that Riley had initiated.

[4] George Marsden, *Fundamentalism and American Culture: the Shaping of Twentieth Century Evangelicalism 1870-1925* (New York: Oxford University Press, 1980), pp. 146-147.

[5] Cited in Daphane Read, ed., *The Great War and Canadian Society: an Oral History* (Toronto: New Hogtown Press, 1978), p. 217.

[6] Marsden, *Fundamentalism and American Culture*, p. 185.

[7] Louis Gaspar, *The Fundamentalist Movement 1930-1956* (Grand Rapids, Michigan: Baker Book House, 1963), p. 77.

[8] Ibid, p. 125.

[9] Hal Lindsey, *The Late, Great Planet Earth* (Grand Rapids, Michigan: Zondervan, 1970).

[10] Margaret Atwood, *The Handmaid's Tale* (Toronto: McClelland-Bantam, 1985).

[11] Harvey Cox, *The Secular City: Urbanization and Secularization in Theological Perspective* (New York: MacMillan, 1965).

[12] Harvey Cox, *Religion in the Secular City: Toward a Postmodern Theology* (New York: Simon and Schuster, 1984).

[13] Jeffrey K. Hadden and Anson Shupe, eds. *Secularization and Fundamentalism Reconsidered*, vol. 3, *Religion and the Public Order* (New York: Paragon House, 1989), p. xxii.

[14] Steve Gotowicki to the Political Islam list, 8 January 1995.

[15] R. Scott Appleby and Martin Marty, eds. "Conclusion" in *Fundamentalisms Observed*, vol. 1 (Chicago: the University of Chicago Press, 1991), p. 816.

[16] Jeffrey K. Hadden and Anson Shupe, eds. *Secularization and Fundamentalism Reconsidered*, vol. 3, *Religion and the Public Order*, p. 111.

[17] Ron Stodghill II, "Religion," *Time*, October 6, 1997.

[18] John D. Spalding, "Bonding in the bleachers: a visit to the Promise Keepers," in *The Christian Century* (March 6, 1996), p. 262.

[19] Bruce Lawrence, *Defenders of God: the Fundamentalist Protest against the Modern Age* (San Francisco: Harper and Row, 1989), p. ix.

[20] *The Globe and Mail* (March 28, 1991).

[21] Peter Marshall and David Manuel, *The Light and the Glory* (Old Tappan, New Jersey: Fleming H. Revell Company, 1977).

[22] Ibid, p. 336.

[23] Martin Marty and R. Scott Appleby, eds. *"Conclusion,"* in *Fundamentalisms and the State: Remaking Politics, Economics, and Militance*, vol. 3 (Chicago: University of Chicago Press, 1993), p. 621.

[24] Lawrence, *Defenders of God*, p. ix.

[25] This attack on Darwin's theory is making a comeback as various "scientists" claim that creation science is as valid a scientific theory as evolutionary theory. Court cases have been initiated, although to date in the United States, the legal results are the opposite of what happened in the famous "Monkey Trial." Even the chief actors seem to be the same – southern fundamentalists versus northern big-city lawyers and liberal theologians. I remember sitting in a classroom listening to the University of Chicago theologian, Langdon Gilkey share his experience as an expert witness. He had been called by lawyers in New York City to testify that creation science was not really science but religion and thus should be prohibited from being taught in American public schools. Langdon Gilkey at that time had long hair and wore an earring in his ear. He recounted that the lawyer from New York, a bright young woman, told him that the earring had to go, as it would hurt that court case. Langdon replied that this was fine with him. The lawyer then continued that he would also need to get a haircut. At this, Langdon Gilkey balked. "Why most of the prophets had long hair," he noted, "and Jesus had long hair and, for all we know, God had long hair." "No *she* didn't!" the lawyer snapped back.

[26] Ernest Sandeen, *The Roots of Fundamentalism: British and American Millenarianism, 1800-1930* (Chicago: University of Chicago Press, 1970).

[27] See Ernest Sandeen, *The Roots of Fundamentalism*, pp. xxi-xxii.

[28] Douglas W. Frank, *Less Than Conquerors: How Evangelicals Entered the Twentieth Century* (Grand Rapids, Michigan: William B. Eerdmans, 1986).

[29] Cited in Randall Balmer, *Mine Eyes Have Seen the Glory: a Journey into the Evangelical Subculture in America*, expanded edition (New York: Oxford University Press, 1993) p. 172.

CHAPTER EIGHT

[1] See Fidel Castro "A Necessary Introduction," in Che Guevara, *The Diary of Che Guevara Bolivia: November 7, 1966 – October 7, 1967*, edited by Robert Scheer (New York: Bantam Books, 1968).

[2] Cited in Deane William Ferm, *Third World Liberation Theologies: an Introductory Survey* (Maryknoll, New York: Orbis Books, 1986), p. 13.

[3] Gustavo Gutiérrez, *A Theology of Liberation: History, Politics and Salvation* (Maryknoll, New York: Orbis Books, 1973).

[4] Juan Luis Segundo, *The Theology of Liberation* (Maryknoll, New York: Orbis Books, 1976).

[5] Part Four, chapter one, section 1134 "Puebla Document," in John Eagleson and Philip Scharper, *Puebla and Beyond: Documentation and Commentary*, translated by John Drury (Maryknoll, New York: Orbis Books, 1979).

[6] Cited in Paul Sigmund, *Liberation Theology at the Crossroads* (New York: Oxford University Press, 1990), p. 121.

[7] Aharon Sapsezian, "Ministry with the Poor: an Introduction," in *International Review of Mission*, LXVI (January 1977), p. 4.

[8] Jon Sobrino, "Preface to the English Edition," in *Mysterium Liberationis: Fundamental Concepts of Liberation Theology*, edited by Ignacio Ellacuría and Jon Sobrino (Maryknoll, New York: Orbis Books, 1993), p. xiii.

[9] Juan José Tamayo, "Reception of the Theology of Liberation," in *Mysterium Liberationis*, p. 36.

[10] Ronald Nash, ed., *On Liberation Theology* (Grand Rapids: Baker book House, 1984), p. 245.

[11] Enrique Dussel, "Theology of Liberation and Marxism," in *Mysterium Liberationis*, p. 88.

[12] Paul Sigmund, *Liberation Theology at the Crossroads*, p. 177.

[13] Cited in Juan José Tamayo, "Reception of the Theology of Liberation," in *Mysterium Liberationis*, p. 43.

[14] Robert McAfee Brown, *Liberation Theology: an Introductory Guide* (Louisville, Kentucky: Westminster/John Knox Press, 1993) pp. 138-139.

[15] Cited in Arthur F. McGovern, *Liberation Theology and Its Critics: Toward an Assessment* (Maryknoll, New York: Orbis books, 1989), p. 227.

[16] Alistair Kee, *Marx and the Failure of Liberation Theology* (London: SCM Press, 1990), p. 257.

[17] Ibid, p. 271.

CHAPTER NINE

[1] Naomi R. Goldenberg, *Changing of the Gods: Feminism and the End of Traditional Religions* (Boston: Beacon Press, 1979), p. 92.

[2] Alistair Kee, *Marx and the Failure of Liberation Theology* (London: SCM Press, 1990), p. 258.

[3] Richard Tarnas, *The Passion of the Western Mind: Understanding the Ideas that Have Shaped our World View* (New York: Ballantine Books, 1991), pp. 441-445.

[4] Anne Carr, *Transforming Grace: Christian Tradition and Women's Experience* (San Francisco: Harper and Row, 1988) p. 11.

[5] Anne Carr and Douglas J. Schuurman, "Religion and Feminism: a Reformist Christian Analysis," in *Religion, Feminism, and the Family,* edited by Anne Carr and Margaret Stewart Van Leeuwen (Louisville, Kentucky: Westminster John Knox Press, 1996), p. 15.

[6] Mary Daly, *Pure Lust: Elemental Feminist Philosophy* (Boston: Beacon Press, 1984), p. 24.

[7] Ibid, p. 2.

[8] Ibid, p. 414.

[9] Daphne Hampson, *Theology and Feminism* (Oxford: Basil Blackwell, 1990), p. 76.

[10] Pamela Dickey-Young, *Feminist Theology/Christian Theology: in Search of Method* (Minneapolis: Fortress Press, 1990), p. 113.

[11] Jacquelyn Grant, *White Women's Christ and Black Women's Jesus: Feminist Christology and Womanist Response* (Atlanta: Scholar's Press, 1989), p. 200.

[12] Elizabeth Schüssler Fiorenza, "Changing the Paradigms," in *How My Mind Has Changed*, edited by James M. Wall and David Heim (Grand Rapids, Michigan: William B. Eerdmans, 1991), p. 78.

[13] Chinnici, *Can Women Re-Image the Church?* (Mawak, New Jersey: Paulist Press, 1992), pp. 95-96.

[14] Rosemary Radford Ruether, *Women-Church: Theology and Practice of Feminist Liturgical Communities* (San Francisco: Harper and Row, 1985), p. ix.

[15] Ibid, p. 137.

[16] Name changed.

[17] Elizabeth A. Johnson, *She Who Is: the Mystery of God in Feminist Theological Discourse* (New York: Crossroad, 1992), p.33.

[18] Marilyn Todd, Internet Letter to Femrel List, (29 November 1994).

[19] Stanley J. Grenz and Roger E. Olson, *20th Century Theology: God and the World in a Transitional Age* (Downer's Grove, Illinois: InterVarsity Press, 1992), p. 234.

[20] Elizabeth Achtemeier, "The Impossible Possibility: Evaluating the Feminist Approach to Bible and Theology," in *Interpretation* 42 (January 1988), p. 57.

[21] Fiorenza, "Changing the Paradigms," p. 80.

[22] Ruether, *Women-Church*, p. 60.

[23] Elizabeth Moltmann-Wendel and Jürgen Moltmann, *Humanity in God* (New York: Pilgrim Press, 1983), p. 111.

CONCLUSION

[1] Robert Heilbroner, *Visions of the Future: the Distant Past, Yesterday, Today, and Tomorrow* (New York: Oxford University Press, 1995), p. 95.

[2] Of course, one might gently point out that while Nicky Cruz may have had a wonderful conversion story, he was unschooled in theology. However, even so formidable a scholar as Romano Guardini – professor of Christian philosophy at the universities of Breslau, Berlin, Tübingen, and Munich, and one of the leading participants in the *ressourcement* movement which culminated in the historic Vatican II Council – fell into the same trap. One of the first to use the term "new age," Guardini wrote a series of letters in which he reflected on the shape of faith both in the past and in the future. In one of these letters, he stated that,

...a new image is in the process of formation, a different one from that of antiquity or that of the Middle Ages, and especially from that of humanism, classicism, or romanticism. It is coordinated with the new events to which we have referred. Further, it is coordinated with the depth of humanity that we hope is coming to the surface. It is coordinated with the new level on which the battle with the forces that have emerged will be fought. And it will be victorious on that level. The new age will be created out of that depth, out of that image.

Guardini's letters were written between the years 1923 and 1925, and many of the young people whom he believed would inaugurate this *new age*, members of the German Catholic Youth Movement, in the not too distant future would be willing participants in the establishment of Adolph Hitler's Third Reich – hardly the future which Guardini envisaged. See Romano Guardini, *Letters from Lake Como: Explorations in Technology and the Human Race*, introduction by Louis Dupré, translated by Geoffrey W. Bromiley (Grand Rapids: Wm. B. Eerdmans, 1994), pp. 89-90.

[3] The biblical prophet Amos' response to a ban on his prophetic activity at Bethel, the chief religious and political center of the northern Kingdom of Israel.

[4] John Naisbitt and Patricia Aburdene, *Megatrends 2000: Ten New Directions for the 1990s* (New York: Avon Books, 1990), p. 290.

[5] *The Confessions of St. Augustine*, Book 1, Chapter 1.

[6] James Baldwin, *The Blues for Mister Charlie* (New York: Dell, 1964), pp. 33-34.

[7] The fascinating thing about Bill Gates is the speed with which he became the world's richest individual, as well as the product that made him rich – ideas rather than land, gold, or oil. Gates is not alone in becoming a multibillionaire due to the growing influence of the computer revolution. A whole new class of wealthy individuals is being created by the computer and Internet revolution, challenging families whose wealth has been passed down from generation to generation and whose world view is conditioned by continuity rather than by abrupt change.

[8] See Alvin Toffler, *The Third Wave* (New York: William Morrow, 1980), p. xxii.

[9] For a few seconds the laws which governed the formation of the universe, according to the Big Bang theory, were quantum rather than deterministic, Newtonian laws.

[10] John Polkinghorne, *Quarks, Chaos and Christianity: Questions to Science and Religion* (New York: Crossroad, 1996), p. 16.

[11] It must be kept in mind that the concept of the resurrection of the dead, especially as outlined by the apostle Paul, is a very different concept from the afterlife implied by NDEs. The former depicts a situation where the dead die but are then reassembled by God at the final resurrection while the latter is based on the concept of an immortal spirit which lives on after the death of the physical body. Nonetheless, in spite of the differences between these two approaches, Frank Tipler makes the interesting observation that,

In the fall of 1990 the annual meeting of the American Academy of Religion happened to be held in New Orleans. I attended the plenary lecture by a famous Columbia University historian of the Middle Ages, who spoke on medieval beliefs about life after death. She discussed at length an analysis by St. Thomas Aquinas, the greatest of the medieval theologians, of a technical problem which arises with the idea of the resurrection of the dead: if the universal resurrection is accomplished by reassembling the original atoms which made up the dead, would it not be logically impossible for God to resurrect cannibals?... The audience, several hundred theologians and religious studies professors, thought this quaint 'problem' hilarious, and laughed loudly.

I didn't laugh. When I first read Aquinas' analysis, which I came across when I first began to consider seriously the technical problems associated with a universal resurrection, I did laugh. But I soon realized that Aquinas' cannibal example was subtly chosen to illustrate the problem of personal identity between the original person and the resurrected person; establishing this identity is the central problem to be solved in any theory of resurrection of the dead... I infer that the typical American theologian/religious studies professor has never seriously thought about the resurrection of the dead. Eschatology has been left to physicists.

Frank Tipler, *The Physics of Immortality: Modern Cosmology, God and the Resurrection of the Dead* (New York: Doubleday, 1994), p. xiii.

[12] Richard Tarnas, *The Passion of the Western Mind: Understanding the Ideas that have Shaped our World View* (New York: Ballantine Books, 1991), p. 730.

[13] Naisbitt and Aburdene, *Megatrends 2000*, p. 295.

[14] Douglas John Hall, *Thinking the Faith: Christian Theology in a North American Context* (Minneapolis: Fortress Press, 1989), p. 449.

[15] Peter Berger, *A Far Glory: the Quest for Faith in the Age of Credulity* (New York: Doubleday, 1992), p. 147.

[16] G. W. F. Hegel, *Lectures on the Philosophy of Religion*, Vol. 1, translated and edited by E. B. Speirs (New York: Humanities Press, 1974), p. 22.

[17] Sam Keen, *Hymns to an Unknown God: Awakening the Spirit in Everyday Life* (New York: Bantam Books, 1994), p. 121.

[18] Ron Graham, *God's Dominion: a Skeptic's Quest* (Toronto: McClelland & Stewart, 1990), p. 394.

[19] James 2:19 (NRSV)

[20] Albert Schweitzer, *The Decay and Restoration of Civilization: the Philosophy of Civilization Part 1*, translated by C. T. Campion, 2nd Edition (London, A & C Black, 1932) p. 97.

[21] Paul Tillich, *Systematic Theology*, Volume 3, *Life and the Spirit, History and the Kingdom of God* (Chicago: The University Press, 1963), p. 250.

[22] T.S. Eliot, *Collected Poems 1909-1962* (London: Faber and Faber, 1963), p. 222.

INDEX